P9-BZG-620

# GLOBAL WARMING

## Can Civilization Survive?

### PAUL BROWN

BLANDFORD

## ACKNOWLEDGEMENTS

Thanks to Tim Radford, Science Editor of the *Guardian*, and Rachael Baird for their encouragement and help in looking out research materials; Merylyn McKenzie Hedger for saving so many documents and discussing the politics; my mother, Kathleen, for reading the first draft and Maureen, my wife.

**A BLANDFORD BOOK**

First published in the UK 1996 by Blandford
A Cassell Imprint
Cassell Plc, Wellington House,
125 Strand, London WC2R 0BB

Copyright © 1996 Paul Brown
The right of Paul Brown to be identified as author of this work has been asserted by him in accordance with the provisions of the UK Copyright, Designs and Patents Act 1988.

All rights reserved. No part of this book may be reproduced or transmitted in any form or by any means, electronic or mechanical, including photocopying, recording or any information storage and retrieval system, without permission in writing from the publisher.

Distributed in the United States by Sterling Publishing Co., Inc.,
387 Park Avenue South, New York, NY 10016-8810

British Library Cataloguing-in-Publication Data
A catalogue entry for this title is available from the British Library

ISBN 0-7137-2602-4

Typeset by Bibliocraft, Dundee

Printed and bound in Great Britain by
Biddles Ltd, Guildford and King's Lynn

# Contents

# Introduction

---

## Can civilization survive the crisis it has created?

THERE IS NO LONGER any scientific doubt that we are causing the earth's climate to change. Global warming is not a distant theory but a fact of our lives. Yet the urgency of the crisis seems hardly to have been noticed by the general public. They may have heard scientists warn of a series of coming catastrophes which large parts of the civilized world may not survive, but somehow it seems not to apply to them. Certainly they have not demanded action from politicians. The scientists, having done their job, send their reports to governments, and go off and do more research. Ministers sometimes receive these predictions with genuine shock, promise to do something, and then let the world go on the same way as before.

But behind the scenes much is stirring. There are serious political rumblings over the issues of climate change, and the outcome promises to be as important as the ideological struggle used to be between Communism and Capitalism. There are still those who argue against action and those who do not accept the consensus that there is a problem. These groups and individuals support those few scientists who still argue that climate change is not happening at all, and provide funds for the sophisticated lobbyists acting for the coal and oil industries and the states that rely on the revenue from fossil fuel production, who see action on climate change as a threat to their future. At the opposite extreme are those who think that action is urgent now, in particular the small coastal and island nations that are likely to disappear below the rising sea and the environment groups whose interest is in keeping these issues in the public eye.

This book brings together the science and the politics to describe the stage we have reached on the questions raised by climate change. We name some of the main players, countries and individuals, heroes or villains, depending on your point of view. Global warming has a short history and a long future, whichever way the world chooses to deal with it. The decisions made in the next few years, and the decisions not made, will dictate the shape of the world to come.

To endeavour to understand the issues is not easy, and so the book has been split into four basic strands. The first is a short history of what has happened, with the use of journalistic hindsight. Having reported on some of the conferences and political events at the time for a daily newspaper, I find it surprising how different they look in retrospect. The recent history shows how the science and the politics became intertwined, making the issue even more difficult to tackle.

The second section is an explanation of what the 'greenhouse effect' is, an update on the latest science of what is happening, what is making the world warm and causing the sea to rise. The third part is the current and future effects: the flooding, the storms, the droughts and famines – and likely mass migrations and wars. This is based on the latest science published in 1996 and pinpoints the areas of the world which will be affected first and worst according to current estimates. Finally, the fourth section deals with the politics, past, present and future, where the main players stand, what their aims appear to be. It reports the struggle for the agreements that will be the basis for future action.

Clearly, international effort is needed to deal with a global problem. First we need meaningful agreements to do something, and then we need to implement them. This will mean hard decisions, and some would say sacrifices, which will affect us all for the sake of our children, or indeed ourselves if we expect to live for another 30 years. These decisions have to be made by politicians whose normal instincts are not to make them, and to put their own electoral survival and narrow national interest at the top of any agenda.

First some basic points on this controversial but fascinating subject. As a journalist covering this issue for ten years I have written many articles on climate change as environment correspondent for the *Guardian* newspaper in London. In pursuit of information the paper has dispatched me to meet many scientists and attend many conferences. These began with the preparatory meetings as politicians and civil servants struggled to understand the issues and decide what to do. Then it was on to the World Climate Conference in 1990, the Earth Summit in Rio in 1992, and the Climate Conferences in Berlin in 1995 and Geneva in 1996. At all levels, I have found that the subject arouses enormous passions. As the situation becomes more critical the power struggle intensifies and it becomes increasingly difficult to write about the subject and express opinions without treading on some toes.

In my researches I came across a writer who compared the attitude of scientists advising governments on the dangers of global warming to that of cold war generals advising on the military threat of the Soviet bloc. He said that if scientists had the same attitude to threats to the planet as generals had about the threat of Communism we would be well on the way to dealing with global warming by now. As it is, we have hardly started. Scientists are convinced by their own research that global warming is a great threat to humanity and yet they have been remarkably quiet. Worried about professional reputations and not wishing to cause a stir they have been content to go back to their universities to do more work. But if they are right about the dangers, and the evidence piles up all the time that they are, then they have been failing in their duty. They need to take a leaf out of the generals' survival manual and voice their concerns about the coming crisis until the message gets across.

What follows is a personal view of what has been done so far, needs to be done in the future, and how we are limping along towards only a few of the necessary targets. Vast volumes of scientific reports have been read, political advice studied, and available books consulted. It is the facts as they appear to me, with my best effort to get them right. That does not mean that they are all correct, or that some that are right will not be challenged. The most important point is that we should confront the problem.

To give a flavour of what follows, the rest of this chapter summarizes the point we have reached in terms of scientific research and political action.

The latest report from scientists published in 1996 says that gases released by the activities of human beings from the industrial

# Global near-surface temperatures 1860-1995

Difference in degrees Celsius from the 1961-90 average

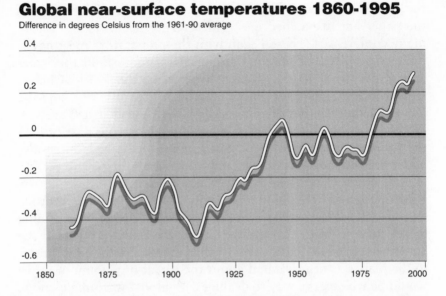

revolution onwards are altering the composition of the atmosphere and causing it to heat up. These are not just any old scientists, they are hundreds of the best authorities on climate, the cream of the world's experts appointed to the United Nations' Intergovernmental Panel on Climate Change (IPCC). In the first of three reports (IPCC I) they agree that the ever-increasing use of coal and oil to fuel a worldwide consumer society will make the process of climate change rapidly worse. They think that the change is happening at a pace so fast that it is outstripping the ability of plants and trees to adapt.

Mountain glaciers are melting, along with ice at the poles. The oceans are getting warmer and expanding. The result is that sea levels are rising. Among the predictions is that sea levels will rise so fast in the next 50 years that some low-lying island nations may disappear altogether. In the unprotected river deltas of Bangladesh, Egypt and Vietnam there are a total of 30 million people living within 1 m (3 ft) of the high-tide mark. They have no safeguard.

Equally horrifying is the fate of others who will go hungry as crops fail. There are 100 million people in ten sub-Saharan African states who are already living on the edge of starvation because of prolonged droughts in 1991–92. These droughts are believed to be one of the first signs of climate change. The rise in the severity and frequency of tropical storms has already bankrupted some insurance companies. This means that many people in storm belts are no longer able to obtain insurance.

The IPCC scientists are saying in their 1996 report on the science of climate change that they can detect some human-induced global warming already and there is worse to come whatever we do. That is because there is a time lag between creating the problem and its becoming apparent. It used to be thought that it takes about 30 years for the climate to catch up with the cocktail of gases we have already pumped into the sky. The scientists are now saying it will take hundreds of years for the temperature of the land surface to peak and even longer for the sea to heat up and reach equilibrium. That means that whatever we do now, sea levels will go on rising for centuries, reducing the land area available for humans to live on.

In other words we cannot stop the process we have started. We have already, unknowingly, put enough extra global warming gases in the atmosphere to keep the process going for a long time yet. Global warming is going to happen anyway, what we need to do is to control it and slow it down, so we have time to get used to it and change our agriculture and our lifestyles. What we need to do, urgently, say the scientists, is to reverse the trend of ever-upward emissions. It is not enough to cut the rate of rise, emissions need to fall progressively to counteract the effect of the extra we have already added, and get back to something like a natural balance.

The first IPCC report to the United Nations in 1990, *Climate Change*, suggested that to return the planet's climate to a manageable state a 60 per cent cut in carbon dioxide emissions was needed. That seems impossible, so a better way of looking at the problem is to say we have to aim at something approaching a 1 per cent reduction of emissions a year. By doing that we can slow the speed of climate change. This gives the human race and the rest of the natural world time to adapt to the problem and avoid a calamity. At the moment, though, emissions are still rising fast.

Already we have all heard of droughts, heatwaves, storms and floods being blamed on global warming or more likely 'the sort of thing we would expect from the greenhouse effect' but they are usually one-off events or disasters which have been recorded before. We now need to ask whether they will become a regular occurrence and whether we can live with them. But that is only one area of questioning.

Scientists have compiled reports running to thousands of pages on the severity of the consequences of climate change and which changes will affect our lives first. But it is still the subject of a great deal of debate. It is also part of the biggest collaborative research programme the world has ever known. New information is being released all the time. Doubts about what is happening are gradually being eliminated. Real observation increasingly matches computer predictions, but the margins for error remain wide.

This vast amount of information is remarkable, considering that climate change is such a relatively recent issue. It arrived on the political agenda as late as 1985. Margaret Thatcher, the British Prime Minister, in 1990 spoke to the United Nations of 'the greatest threat facing the human race'. Al Gore, later to become vice-president of the United States, wrote a book about it. In fact when world history comes to be recalled at the end of the next century the ten years from 1985 may not be remembered chiefly for the end of Communism and the collapse of the Berlin Wall but as the time humans first realized that they were destroying their own environment. Perhaps the politicians will be remembered for taking the first historic steps to do something about it.

In these last ten years a whole range of worldwide threats have been identified and acknowledged as requiring action. Some, for example the hole in the ozone layer, have been tackled convincingly, so there is hope, not just of staving off disaster, but for a new era of international understanding based on a common wish to survive.

Getting action on climate change is not going to be as easy as it was with ozone. Some believe the problems are so difficult it is not even possible to solve them. But a high spot, which shows what can be done with enough political will, was the Earth Summit in Rio in Brazil in 1992. A gathering of world leaders, in the largest international conference ever organized, signed a number of important international agreements to help the environment. Among them was the Climate Change Convention.

At the time many people wrote off the Earth Summit as a failure, and indeed there are many pledges made in Rio that remain to be honoured, not least on climate. Nevertheless, it began a political process which has developed its own momentum and provides the opportunity to tackle the coming crisis. The latest policy documents show how far thinking has developed since then. They show that much can be done to tackle the problem at little or no cost to the economic welfare of individual countries. In fact they are the things we should be doing already to conserve natural resources. Some of them are already underway in a minor fashion. These include recycling, increasing energy efficiency in machines, insulating buildings and switching to renewable energy.

Meanwhile in the real world outside the conference centres the volumes of greenhouse gases pumped into the atmosphere increase every day, and the developing world is only just undergoing its industrial revolution. We have already seen wars about oil. The next round of wars are expected to be about water, an even more vital resource. In parts of the parched Middle East the main rivers often flow through more than one country. Thus an irrigation supply used with profligacy for one country can evaporate away the drinking water and life support system for the countries downstream. Where rivers continue to flow, agreement is possible over sharing the water. When the rivers dry up altogether the consequences are disastrous. Massive migrations of people are expected to be one of the first results of climate change.

But as the Earth Summit made clear, global warming is only part of a wider picture of related problems that have to be tackled. Cutting down forests changes the climate. We have been doing this for thousands of years. In the Mediterranean region the Romans realized too late that they had turned green fertile lands in North Africa into arid zones by cutting down the woodlands. It clearly takes a long time for human beings to learn by their own mistakes. Today we realize that overintensive agriculture in vast areas of the modern world is creating or extending deserts and yet we carry on. All over the globe people's activities are wiping out species. The planet's precious diversity is being damaged each day.

You could say that at least we understand what we are doing and have the knowledge at our disposal to do something about it. At any moment, somewhere in the world, there is a conference going on trying to tackle some of these problems, or related ones, like population growth. Progress is being made but nearly every one of these world problems depends on joint action. Sometimes the very fact that more than one nation needs to take part even to make a dent in a problem is used by politicians as an excuse for doing nothing. And ministers have not made it any easier on themselves by pretending to be 'green'. For example, in Britain the government decided to add 15 per cent value added tax to gas and electricity bills to reduce consumption. There was anger because the public did not believe the tax was to prevent global warming as the government claimed, they thought merely that the Treasury saw it as an easy way to make a fast buck. The move was rejected by Members of Parliament. This will make it more difficult to impose green taxes, even legitimate ones, in the future.

In the end there will be no substitute for concerted international action through agreements. In the case of global warming the issue seems straightforward. What is needed are targets for reducing emissions and timetables for these targets to be reached. But this simple answer requires a series of hard political decisions on the domestic front if politicians are to honour their pledges. As a result, getting agreement on the international stage has proved difficult. This realization has meant that since the Earth Summit progress has been painfully slow, and new international commitments have not been made. The next big leap forward is due to be in 1997 when targets and timetables must be agreed if real progress is to be made beyond the year 2000.

It is easy to become depressed by the lack of progress, but time and time again at these conferences when all has appeared to have been lost there is an unexpected political surge. This happened at the Berlin Climate Conference in 1995 on the very last day of talks. Politicians, through their officials at negotiations, seemed capable of stalling for months, but faced with public failure and the resultant bad publicity of going home empty-handed a deal was suddenly agreed at the last moment.

It is also easy to be cynical, but it is occasionally refreshing to see that some politicians do really think about what they are doing, and a few have studied and understood the evidence about global warming. One such is John Gummer, who as environment secretary was the UK representative in Berlin. Later in this book there will be some unkind things said about the long years of the Conservative government and its environmental record, but Mr Gummer certainly understands the threat of global warming. He mentions it repeatedly in speeches because he is clearly alarmed by the prospect. The following is an extract from an after-lunch speech he made in January 1996 to the Royal Institute for International Relations in Brussels:

> I am deeply concerned that we no longer understand, as we did understand in Rio, that global warming is not something which

can be discussed and dismissed; it's something that needs to be acted upon. That we need now, not tomorrow, but now, to take measures which alone can save our children from a degree of climate change which would be instantly devastating. When we talk about our children we talk about the vision of Rio. Then we understand that it was not children in a general way – the future generations – about whom we spoke, but in a very specific way. We were saying that within the working lifetime of the children of people here present, the climate will change to a degree which will utterly alter the basis of their lives.

Mr Gummer was talking from the heart and he is never prone to exaggeration. He had seen the latest scientific evidence, and in my view he is right to keep sounding the alarm. It is disappointing that he and other concerned politicians have been unable to carry their political colleagues along with them.

The conclusion I come to is that there are two possible consequences of climate change. The first is what we could call the 'nightmare scenario'. It runs something like this:

Scientists have told us that global warming has already set in. This means that more and more extreme weather events become a feature of our lives. Droughts, floods, heatwaves, and sudden cold snaps damage homes and businesses. Agriculture is disrupted and world food supplies are inadequate. Mass migrations of near-starving people begin in Africa and the Middle East, bringing new wars. A large wave of bankruptcies in the industrialized world saps the economic ability of the so-called developed world to deal with the problem. There is a worldwide recession while politicians attempt to grapple with domestic problems and the root cause of the problem, the destruction of the environment, is not tackled. In the developing world the new industries which have dragged millions of people from the countryside into the cities lose their markets. The recessions throw millions out of work in countries that have no social security systems. Civil unrest and a few revolutions follow. The world begins to lapse into anarchy, making it impossible to deal with climate change. Meanwhile the gases we have already put into the atmosphere go on making things worse. The rising sea level means some of the most fertile and populous land is inundated, driving the survivors inland to find new homes. Deserts continue to expand; rainfall has become unreliable in marginal lands. Civilization breaks down and millions, possibly billions, die.

In the second scenario, the politicians come to grips with the problem. They grab the opportunity in the next two years to cut emissions of greenhouse gases and start setting targets and time-tables to do so from 2005. This convinces the developing world that the industrialized world, which caused the problem in the first place, is serious about making amends. Money is poured into research to find new technologies to generate electricity without producing carbon dioxide. This technology is offered to the poorer countries so they can reduce their own emissions while at the same time allowing development to improve the lot of their people. By the middle of the twenty-first century greenhouse gas production has dropped by 60 per cent on current emissions. The human race is beginning to adapt to the climate changes that are taking place from past emissions. The way of life of billions of people has already changed. Commuting to work in a car has been confined to history as an out-of-date and distinctly anti-social activity. Cars that use fossil fuels are only seen in museums and those on the road are driven by hydrogen or some other non-polluting fuel. Most people have ready access to a rapid transit system and do not need a car. Out-of-town shopping centres, which need a car for access, have long since become white elephants and been closed down. Activities which damage the environment are seen as anti-social as civilization is at last educated to appreciate that without a thriving natural world the human race is doomed.

At present we are poised between these two extremes. Mr Gummer has looked over the edge into the abyss and does not like it, but at present in all the large democracies the short-term politics of winning the next election and the need to increase the annual profits of industry rule over the long-term interests of the human race. But human beings also have vision and ingenuity to understand and overcome problems. Which of these traits will triumph? In the rest of the book we consider the progress towards success or failure in this battle for the survival of civilization – the greatest test humanity has ever faced.

# PART ONE

---

# THE HISTORY

# 1

## When we started to worry

CONSIDERING THAT CLIMATE CHANGE is such a great threat to the future of
the human race, scientists and politicians have been worrying about
it for a remarkably short time. In the 1970s when scientists were
beginning to grapple with the complexities of the theory that our
activities might be causing the climate to heat up, there was a great
deal more publicity about the possibility of a new ice age because
there had been a couple of cold winters. In the 1980s when scientists
first began to worry about the world beginning to overheat, it took
some time for them to attract the attention of public and politicians,
and then it was only when we experienced some unusual weather
that they were taken seriously.

Not that the science was new. The basic principles were first noted
in 1827 by a French mathematician and scientist, Jean Baptiste
Fourier. He observed that certain gases trapped heat in the atmosphere and he coined the phrase the 'greenhouse effect', displaying a
rare knack, for a scientist, of describing a complex process so simply
that everyone could understand it. He said that the atmosphere was
like the glass in a greenhouse. It let the sun's rays in and therefore
the warmth but it also provided a barrier which prevented the
accumulated heat escaping again. Some gases in the atmosphere,
he explained, particularly carbon dioxide, acted like those panes of
glass, forming a barrier to the heat getting back into space.

A British scientist, John Tyndall, developed the theory in 1860
by measuring the absorption of heat by carbon dioxide and water

vapour. He hypothesized that a reduction in the amount of carbon dioxide in the atmosphere and therefore a decrease in the greenhouse effect might have been the cause of the ice ages. Another 36 years passed before Swedish chemist Svante Arrhenius considered, and attempted to measure, the possibility of significant increases in carbon dioxide in the atmosphere – exactly the problem that now faces the world. He believed that doubling the concentration of carbon dioxide would increase the temperature by 5–6°C (41–43°F). This is remarkably close to scientists' current predictions, despite all the variables that have now been built in.

All this theory was picked up in 1938 by G. S. Callender, a British meteorologist, who tried to persuade a sceptical Royal Society that global warming was already taking effect. He had gathered information from 200 weather stations around the world and demonstrated that average temperatures had increased between the 1880s and the 1930s. His theory that this was caused by increased carbon dioxide in the atmosphere was greeted without enthusiasm.

He proved to be well ahead of his time and it was not until 1957, when two scientists from the Scripps Institute of Oceanography in California warned that the human race was carrying out a huge experiment with the atmosphere of the entire planet, that the scientific community began to take the issue seriously. Roger Revelle and Hans Suess thought that the build-up of carbon dioxide could be dangerous and as a result of these warnings routine measurements of carbon dioxide in the atmosphere were begun at the remote observatory on Mauna Kea, in Hawaii, 3300 m (11,000 ft) above sea level. It was felt that this site, far removed from any industrial source of the gas, would provide a true reading of the contents of the atmosphere. These have helped ever since to give an accurate and, indeed, alarming graph of the rapid build-up of carbon dioxide in the air since that initial warning.

There needs to be a pause in this history at this point, before the science becomes intertwined with the politics, to point out that the terms 'climate change', 'global warming', and 'greenhouse effect', which have been used as meaning the same thing so far in this book, mean different things to different people, particularly the main players in the debate. There are two forms of greenhouse effect. There is the 'natural' greenhouse effect, which is how warm the earth would be if humanity was not changing the climate, and the 'enhanced' greenhouse effect which is the extra warmth caused by the additional carbon dioxide and other gases being introduced. 'Global warming', the popular term, which seems to mean the same thing as the enhanced greenhouse effect, rapidly became a banned phrase in United Nations documents because it was felt to be politically sensitive. It implied that global warming was a proven fact, and some governments could not accept, for political reasons, that it was – otherwise presumably they would have had to agree to do something about it.

This is where the term 'climate change' comes in. Scientists use this phrase because their research shows that while most of the world warms because of climate change, some areas of it – including, apparently, the United Kingdom – may actually cool. In any event politicians rapidly grabbed hold of the term climate change, finding it politically neutral and therefore preferable to global warming. In 1996, however, when scientists confirmed that there *was* an enhanced greenhouse effect, i.e. artificial global warming, all this arguing over terminology was exposed for the charade that it is. The confirmation will not stop the politicians fiddling over words but this explanation will help outsiders to understand how much time has already been wasted while the world has begun to burn.

While we are pausing, it is worth mentioning one of the ironies of looking at climate change history in the light of later events – the crucial role played by the United States. The US was the first country to take the threat of climate change seriously. It also became apparent fairly early when the problem was diagnosed that it was the United States that produced most carbon dioxide per head, and most needed to mend its ways. But in the late 1980s the United States was still pushing the science forward and creating the political organization to carry necessary action when required. By 1990 when the rest of the world finally woke up to the enormity and urgency of the threat the United States had got cold feet. The political and the financial implications to the Americans' fabled energy- and gas-guzzling society must have caused panic in political circles. Since then the United States has been the country which has most prominently dragged its feet in taking political action, and has appeared to pass the baton of leadership to Europe. Although leadership by a committee, which is what Europe often amounts to, famously does not work, the European Union has so far acquitted itself remarkably well.

14

But back to the history. It was in 1965 that the White House first ordered a study into the burning of fossil fuels to see if it could be related to the increases in rising carbon dioxide levels. Five years later at the Massachusetts Institute of Technology there was a symposium to investigate such issues. By 1970 there was enough international scientific and political interest in the subject for the UN secretary general's environment report for that year to refer to the potential for 'a catastrophic warming effect' on the atmosphere.

This was the birth of the politics of global warming; the advent of the idea that the world in general somehow had responsibility for controlling the quality of the air so that we did not damage our own climate. Perhaps because it was such an enormously difficult thought, however, it passed by almost without comment. No one seemed to have a clue how to tackle such an immense issue. It has to be remembered, of course, that at that time caring for the environment was still a novel idea. Although there had been international agreements in the past to protect such things as migrating birds and big game in Africa, the 1970s was the first decade when the environment came to the fore as an international issue. Suddenly the focus shifted from saving wildlife to the pollution created by industry and the overuse of resources like forests. The new concern found its full expression in 1972 at the UN Conference on the Human Environment in Stockholm.

This became the largest conference in the world with 114 countries and 1200 delegates. The delegates focused this growing international concern and debated a huge range of issues. Nevertheless, global warming hardly rated a mention. This was despite the fact that the first international meeting of scientists to discuss long-term climate change had taken place in Sweden the year before. The science was clearly not sufficiently developed and the subject so enormous and complex it defeated politicians' ability to make meaningful statements about it in Stockholm a year later. The conference was much stronger on subjects like marine pollution, for example oil tanker spills, depletion of natural resources and population growth. But the conference was remarkable in another respect; it marked the emergence of a united front by the developing countries which remains very much intact now. Although at the time it was not directly related to the issue of global warming, it is central to the argument.

In 1972 the views of the developing nations were presented as the Founex Report, produced by a group of developing country scientists. Its theme went much like this: pollution is mostly caused by high levels of industrialization, past and present, and this has been caused principally by the developed countries of Europe, North America and Japan. The central environmental problems of developing countries stem from poverty, disease, hunger and exposure to natural disasters. The solution to these environmental problems is more development not less.

It was therefore the responsibility of the developed countries to find a cure for their pollution and to step up the aid and technology transfer to poorer countries so they could solve their own rather different problems.

Thus the North–South divide in environmental matters, as in many others, such as trade, was evident from the beginning. The Stockholm conference also set an organizational precedent that was to be repeated at its monster successor, the so-called Earth Summit, 20 years later. In order to provide agreed texts which politicians could sign up to at the big event there were a large number of preparatory meetings. These were attended by large numbers of journalists who wrote numerous articles worldwide which raised both public awareness and expectations. And refreshingly, a large number of non-government organizations were invited to give their views. Initially these were the environmental groups, Friends of the Earth, World Wide Fund for Nature and the like, but later what could be described as anti-environment groups, like the free marketeer industrialists opposed to any kind of regulation to protect the earth, were also included. They reasoned correctly that they had the right to lobby too.

This big conference in Stockholm produced a text about the world's environment problems which, although worthy, had little in the way of concrete commitments. It did, however, lead to the setting up of the United Nations Environment Programme (UNEP), based in Nairobi. As is the way with such conferences, the political head of steam built up beforehand evaporated away the moment it ended. However, the problems it discussed did not.

By now, academic research into the problems of climate change was going on in many places and the World Meteorological Organization (WMO) sponsored a conference on Long-Term Climate Fluctuations at Norwich in England in 1975. The University of East Anglia is one of the foremost research organizations into global warming and has made a significant contribution ever since to alerting the world to the dangers. In 1975 though, the National Academy of Sciences in the United States was still the front runner. It published a disturbing report, *Understanding Climate Change: A programme for action*, on the possibilities of the warming effect of industrial activities. Like all reports on the subject, before and since, it called for more research, in other words more work for scientists. Two years later it produced another report, this time warning that the implications of projected climate change 'warrant prompt action'.

By 1979 the stage had been set for the first World Climate Conference in Geneva, the scene of many a meeting since on this issue. It is important to note that here, and in the years immediately before and afterwards, it was the scientists and not the politicians that were tackling the problem. It was the World Meteorological Organization which set up the conference, provided the umbrella for further research and organized the follow-up meetings to discuss the findings. In 1979 the scientists noted the increased proportion of carbon dioxide in the atmosphere and put it down to increased fossil fuel burning and deforestation. In other words the burning of trees was releasing the stored carbon in the wood back into the atmosphere. They also described as 'plausible' the theory that this would lead to gradual warming of the lower atmosphere.

A follow-up conference was held in Villach, Austria in November 1980 and two years later the US National Academy of Sciences put some figures on the problem. It said that a doubling of the carbon dioxide concentrations in the atmosphere would cause global warming of between 1.5°C and 4.5°C (35–40°F).

What is interesting about these assessments, with the aid of journalistic hindsight, is the remarkable correspondence between the forecasts and current predictions. Further conferences from 1985 up to 1987 reinforced the previous findings. There were meetings in Villach and Bellagio in Italy at which a substantial body of the world's scientists agreed that global warming was a serious possibility. They offered an estimate of 0.3°C (32.5°F) a decade for the speed of the warming, slightly on the high side of current best estimates but not outside the range of possibilities. The 1987 meeting suggested an international treaty to cut back on the expected rate of release of greenhouse gases to reduce the dangers of excess warming.

The following year, 1988, the WMO and UNEP governing councils decided to set up an intergovernmental organization to assess the scientific information and to formulate response strategies. This became the Intergovernmental Panel on Climate Change, the IPCC. This organization has become immensely important in both the science and politics of global warming. It held its first meeting in November 1988 in Geneva. Its potential importance had not been missed by the major nations and they made sure that their best and most influential scientists were appointed to it.

The IPCC was split into three working groups. The first was to assess the available scientific information. This meant looking at the gases which might cause global warming, assessing whether it was actually happening and by how much, and predicting what was going to happen in the future, including any effects on sea levels. A timetable for the speed of these changes was also required. The second group was to assess the impacts of these expected climate events on the world in general and as far as possible on individual regions. Did it mean more deserts and floods? What was likely to be the effect on health, energy and water resources? The third group was to look at policy, in particular how to mitigate the impact by burning less fossil fuels, for example, and planting forests to absorb carbon dioxide, but also to suggest ways in which the human race could adapt to the changes.

But policy means politics. Scientists may suggest ways of tackling global warming but any changes needed to be implemented by politicians. Science and politics had begun to mix and would not be separated again.

The birth of the IPCC and its assumption of a political as well as a scientific role was not taking place in isolation. The great leap forward into the political arena had begun in June 1988 in the United States, when James Hansen, from the National Aeronautics and Space Administration (NASA), created the first public sensation about the threat of climate change. He said loud and clear for all America to hear that he was 99 per cent certain that the warming of the 1980s was not a chance event but causally linked to global warming. His testimony on television to the US Senate Energy Committee went further than any previous assertion by scientists and was made at the moment of maximum impact. He said it was 'time to stop waffling so much and say the evidence is pretty strong that the greenhouse effect is here.'

This was a key moment in the United States. The summer of 1988 saw a catastrophic drought in the American Midwest which could now be blamed on global warming. Elsewhere in the world there had been a freak hurricane in the English Channel in October 1987 which had knocked down a million trees in southern England alone and caused widespread destruction. In the Antarctic an iceberg 40 km across and 160 km long (25 miles by 99 miles) broke off, leading to discussion about sea level rise and fears that this might be the first sign. Hurricane Gilbert caused widespread damage in the Caribbean.

In June 1988 a conference was held in Toronto called 'The Changing Atmosphere: Implications for Global Security'. By later standards this was a small affair with only 48 countries taking part, but it agreed on what has been known ever since in the jargon of the conference world as the 'Toronto Target'. The conference called for a 20 per cent cut in global carbon dioxide emissions from 1988 levels by 2005 with the eventual aim of a 50 per cent reduction. It also wanted less deforestation, and the establishment of a world atmosphere fund to be paid for by a levy on fossil fuel consumption in developed countries. This was the first mention of a 'carbon tax'. None of the targets adopted by the conference has yet been attained, and of course they were not binding.

Again it shows how little has changed. These are the sorts of targets that those serious about tackling climate change are still demanding. The debate was about what needed to be done rather than what could be done. The realities of what such radical measures might mean in terms of domestic policy have terrified politicians ever since. Tackling climate change means radical policy measures which the politicians think will inevitably lose them a lot of votes. It is worth noting the conference view that humanity 'is conducting an unintended, uncontrolled, globally pervasive experiment whose ultimate consequences could be second only to a global nuclear war'.

Another significant event in 1988, this time in September, was the meeting of the South Pacific Forum, which also discussed climate change. Many of the Forum members are small island states based on coral atolls, the highest points of which are only a few feet above sea level. Their very existence is threatened by the predicted sea level rise. Even if the islands themselves survive, the prospect is that ever more violent storms will wash over them, rendering most of their tiny territories uninhabitable. The Forum predicted that if global warming became a reality its members would produce 500,000 environmental refugees.

Out of this meeting was born the Alliance of Small Island States (AOSIS) which has become one of the important power blocs at subsequent climate talks. In a world dominated by big powers its influence would seem to be limited but its support is far wider than its membership would indicate. Many of the larger Third World countries have vast low-lying deltas which are equally vulnerable to sea level rise. In fact many of these low-lying areas are the most densely populated and productive parts of the countries concerned. The small island states are also heavily supported by the lobbying power of the international green movement.

In this situation it is hard for any politician to argue that nothing should be done about climate change. After all, ministers are sitting in a conference hall with delegates from another country which will simply disappear if no action is taken. AOSIS also acts as a reminder to the developing world that the first victims of climate change will be among their number. It is difficult for them to say that they want to get on with industrialization and ignore global warming when the first such obvious victims of such a policy are on their side and in their midst.

# 2

---

# Ozone:
# a blueprint to save the climate?

CLIMATE CHANGE ALSO MOVED on to the main political agenda in Britain
in 1988 with the surprising intervention of the Prime Minister,
Margaret Thatcher. In a speech to the Royal Society in September
she talked powerfully about the issues of acid rain, damage to the
ozone layer and global warming. From a politician whose preoccupation seemed always to be economics and privatization this came
as something of a surprise, especially as in 1985 she had characterized environmentalists as 'the enemy within'. But Mrs Thatcher
had always prided herself on being a scientist and she had clearly
taken the trouble to explore the subject. Her speech was not just a
one-off either. The following week Sir Geoffrey Howe, who, as a
former chancellor and then foreign secretary, had never apparently
had a green thought, told the United Nations General Assembly:
'We are totally dependent on climate. Damage it beyond repair and
the earth becomes a lifeless desert, spinning in space. We cannot
leave a problem of this magnitude to technical bodies.'

What had caused Mrs Thatcher suddenly to take on the environment as a political issue is not known. Some claim it was that she
read the testimony of James Hansen to the US Senate and became
convinced by it, but by now information on global warming was
coming from all quarters. The House of Commons Select Committee on the Environment had taken similar evidence to that given by
James Hansen but had not delivered it in quite such a dramatic or
high-profile way. It seems most likely that Mrs Thatcher's change of

heart was brought about as a result of the gut instinct of a politician about voters' concerns. That period was a time of economic boom in Britain, a false dawn as it turned out, but at a time of economic prosperity environmental concern shoots up the political agenda. Environment groups like Greenpeace and Friends of the Earth had grown from tiny organizations into powerful lobbying forces in a few short years. Issues like pollution in the North Sea and acid rain had become the cause of much European concern, both to these environmental groups and to governments like West Germany. The British attitude had been always to drag its feet because action on the environment required regulation and interference in the market, two things Mrs Thatcher automatically opposed. For example the British government was the last to acknowledge that acid rain damage in Scandinavia had any relationship to coal-fired power stations, least of all to British ones. Britain had already been dubbed the 'Dirty Man of Europe', an insult that stuck long after it was no longer true. Mrs Thatcher must have paused to find out what the threat from the environment meant in terms of votes, and discovered in passing that there were some pretty frightening facts in view.

While it is possible to overestimate the impact of the intervention of one politician or one speech, Mrs Thatcher's interest took news coverage of the environment from the inside pages to the front of the heavyweight broadsheets. She was also enjoying her highest profile worldwide at the time and was much admired by other world leaders. If Mrs Thatcher thought something was important then everyone took some note.

In November 1988 the first meeting of the IPCC had taken place and by December events were moving fast. A United Nations General Assembly charged WMO and UNEP to make recommendations on the impact of adverse climate change. Perhaps most important, it asked for guidance on a possible future international convention on climate. The political process which led to the Framework Convention on Climate Change had begun. As a result of these decisions the issue maintained its high profile right up to the Earth Summit nearly four years later. A succession of bruising negotiating sessions both on the science and the politics ensured that there was plenty for the world's press to write about.

But before those eventful years are discussed it is a good moment to consider the impact of another important environmental threat

– ozone depletion. Our destruction of the ozone layer was another recently discovered disaster in the making.

Alarms about ozone began a few years earlier than the fears about global warming and its progress as a scientific and then a political issue was always one step ahead of climate change. The discovery of an 'ozone hole' over the Antarctic in the spring of 1985 had galvanized the politicians into action.

Ozone assumes such importance in the climate change story because there had never previously been a problem which demanded such comprehensive worldwide attention. The actions of any one country could wreck the good intentions of everyone else. Action over ozone was, and still is, frequently referred to as a blueprint for action on any other global environmental threat.

The ozone story begins in the nineteenth century with its discovery as a gas in air near ground level. Later in the same century it was thought to also exist high in the atmosphere but it was not until 30 years ago that people began to worry about it. By then ozone was known to shield the earth and therefore human beings from some of the more dangerous ultraviolet light from the sun. At first it was the proposal to build supersonic transport planes that made people worry about whether the exhausts from such inventions would damage the ozone layer. This in itself must have been a first for humanity. Normally we build things and worry about the consequences later. In the event the planes did not materialize but in 1974 the first scientific paper suggesting the theory that chlorofluorocarbons (CFCs) could destroy significant amounts of stratospheric ozone was published. (There is more about CFCs and their role in ozone destruction and global warming in the next section.)

This paper got the international science community moving. Again, the United States led the field and the potential damage from CFCs was intensely examined. In 1976 the United Nations Environment Programme (UNEP) governing council acknowledged the problem and set up an international meeting to discuss what to do. The following year the United States took the lead and hosted the first intergovernmental meeting in Washington to discuss regulating CFC use. The first limit on use came in October 1978 when the United States banned CFCs in non-essential aerosols. Two years later the European Community, having reduced aerosol use by 30 per cent, prevented further production increases.

From then on for five years, a bit like the scientific debate on global warming, the process stuttered along. New bits of science appeared, confirming that damage to the ozone layer was happening, and there were political calls for greater reductions or even bans. The Nordic countries, often the first in the field on environment questions, called for a worldwide ban on CFCs in aerosols and limitations on all uses. A Vienna Convention for the Protection of the Ozone Layer was opened for signature in March 1985, but the political will was still missing.

The final catalyst for action was the discovery by three British Antarctic Survey scientists, Joe Farman, Brian Gardiner and Jonathan Shanklin, of a 'hole' in the ozone layer over the South Pole in the spring of 1985. Unusually for a scientist, Joe Farman did not just write to the scientific magazine *Nature*, record the discovery, and then sink back into more research, he became genuinely alarmed at the significance of his findings and decided that further action was needed. First he wrote to Mrs Thatcher expressing his concern, as one scientist to another, then later he even went on the campaign trail, speaking at press conferences urging action.

The discovery meant another sudden upsurge in publicity; a hole in the earth's shield was a concept that both journalists and their readers understood. The high profile of the campaign increased both the scientific effort and the political will to do something about the problem.

In a reverse of later developments over global warming it was the United States which was pressing for change and Europe and Japan which were rather reluctant to phase out a useful set of chemicals. Dupont, the US industrial giant, could see the writing on the wall and was spending a great deal of money looking for substitutes. By the following year resistance by industry to changing its ways was crumbling. The CFC manufacturers said that safe substitutes for the chemicals might be possible if the price was right, and support from industrial groups came for limits on production. By December legal and technical experts were meeting to draft what was to become a binding international agreement on ozone – the Montreal Protocol. By September 1987, enormously fast by standards of international agreements, the protocol was open for signature.

One of the curious aspects of this protocol was that it called for a cut from the levels of use of CFCs in 1986 of 50 per cent by the year 2000. This was the sort of compromise that could only be reached by politicians. Clearly if CFCs were really destroying the ozone layer then only 100 per cent phase-out was credible. However, politics is the art of the possible and so that was the best that could be done in the circumstances.

The agreement was also only signed by 24 countries, admittedly the main producers and users of CFCs at the time, but it meant that the rest of the world needed to be roped in for the convention to have any meaning. Other similarities between this and subsequent climate change negotiations are worth noting. The first is that the scientific research and meetings to identify and agree on the size of the problem were running in parallel with the political meetings. Lack of scientific certainty was used by some politicians to avoid action for as long as possible. At the same time each new scientific discovery gave the political process an extra shove. But the problems of proving the science were complex and as in climate change the first political agreements to deal with CFCs were in place before there was complete scientific agreement that these chemicals were responsible for ozone depletion.

A second similarity was the involvement of the developing world, which at first was marginal. The key meetings to thrash out the 50 per cent compromise excluded the countries of the 'South' but everyone had realized from the beginning that without their rapid inclusion the agreement was worthless. While China and India then used only a tiny percentage of the world consumption of CFCs, their potential use was enormous. To attract them in to the negotiations a provision was made to allow an extra ten years of use of CFCs before a total ban. It was a trick later to be used for climate change. The issue of how to bring in the developing world and get them to share responsibility for future environmental problems has been carried forward through every international negotiating session since.

The argument of the developing nations at Montreal and through subsequent negotiations went like this: if the industrial world wanted them to help save the ozone layer then it must be at no extra cost to the developing world. The CFC substitutes must be made available when they were developed, but it was not enough to just offer them for sale. The developing world must be allowed to manufacture the substitutes and use the technology and the know-how at no greater cost than if CFCs were still available. This was a pretty tough demand for the industrialized countries to accept but again it boiled down to a simple point – you caused the problem of ozone depletion, you have to pay for it to be fixed.

The third similarity between the issues of ozone and climate is related to the previous two but has particular relevance to what is happening over global warming as we approach the end of the century. It is called the 'ratchet effect'. Once an agreement is reached it allows progress in only one direction – towards an even tighter agreement. By 1990 when a London ministerial conference on ozone took place, the ratchet had turned through a series of meetings. By this time the science had hardened, and each spring the ozone hole over the Antarctic was bigger. It was agreed among other things that the production and use of CFCs should be phased out completely by the year 2000. A group of 13 countries agreed to phase them out by 1997, now that new technologies were available. There was also an agreement to set up a $240 million fund to help the Third World countries obtain the alternative technology and set up their own plants. Of this, two lots of $40 million were set aside for the key Third World players, India and China.

Ozone was a copybook example of the ratchet effect because the dates and deadlines were constantly revised and brought forward. The impending ban had forced industry to look for alternative technologies. They rapidly learned they could make as much money out of new products as selling the old, a lesson still to be fully appreciated on global warming. But it meant that as far as CFC substitutes were concerned that industry got a boost rather than the slap in the face it had originally feared. Another parallel with climate change is that CFCs were found not to be the only ozone depleters. A number of widely used industrial products had the same unfortunate properties. All of them were annexed to the original protocol and a variety of controls and phase-out dates fixed. This is clearly a process that will continue but in 1995 it was reported for the first time that the quantity of CFCs in the atmosphere had started to go down. The ozone thinning is expected to go on getting worse until the end of the century but then to get better. It should have returned to normal by 2050, that is if everyone remains vigilant.

The Montreal Protocol, with its ratchet effect and its provisions for bringing in the developing world, was held up as a blueprint for all future world environmental agreements. Whenever environmentalists complain that not enough has been done and that agreements are too weak, the optimists point to the achievements of the Montreal Protocol. The pessimists say that by comparison with any other problem-mending the ozone hole was simple. There were relatively few sources of CFCs and once alternatives had been found, CFC manufacturing could stop, and the problem was solved.

The reality is, as ever, somewhere between. At the beginning of the ozone negotiations few believed it was possible to get agreement to phase out what had become a staple item for fridges and air-conditioning. Yet in a remarkably short time a total phase-out was agreed and the infringements were few. Even with the example of ozone before them the climate negotiations proved to be far more difficult. It was not that there was a lack of scientific endeavour or political effort. If anything the feeling of urgency was greater because the Earth Summit was looming and the agreement had a deadline to meet. Once the enormity of the problem became apparent, the United States, once driving force behind the whole debate, got cold feet. Conquering global warming was clearly going to be tough.

# 3

## The United States (and the UK) change their spots

THE CHANGE OF HEART by the United States in pushing ahead with action on climate change was not immediately apparent. Huge amounts of money were being poured into research and American scientists were taking a leading role in the Intergovernmental Panel on Climate Change (IPCC). It was no accident that the chairman of the first working group on the pure science was British, the second, on effects, was a Russian and the third, the politically important bit about how to deal with the consequences, was an American.

The Americans had realized the magnitude of the problem early on, particularly from their own domestic political point of view. The United States was the epitome of the consumer society, familiar throughout the world through the eye of the television camera and films. Americans had big cars and every modern electrical appliance, central heating and air-conditioning were normal rather than luxuries. But translate these into volumes of greenhouse gases released into the atmosphere and the results are staggering. At the time of the Earth Summit *Time Magazine* put it in perspective. The United States, it said, has 5 per cent of the world's population, uses 25 per cent of the world's energy and emits 22 per cent of the carbon dioxide. It was also the richest nation, producing a quarter of the world's gross national product. India, by comparison, had 16 per cent of the world's population, used 3 per cent of the world's energy, emitted 3 per cent of the carbon dioxide and accounted for only 1 per cent of the world's gross national product. These figures added force to the Third World argument that they were not the cause of the problem, either historically or in the present.

But most important, the size of the problem and the difficulties of dealing with it must have sent shivers down the spine of every politician in the United States. It was one thing to stop production of CFCs, however expensive, to save the ozone layer, but it was quite another to attack the lifestyle of every American voter just to slow down global warming. After all, the threat of climate change was remote compared with the threat of losing the next election. Although both sets of figures are so large as to be almost incomprehensible, the cost of solving the ozone problem and the cost of tackling global warming were in different leagues. Worldwide it was estimated that the CFC industry was worth $5 billion dollars a year. Phasing it out was to take several years but despite the difficulties the chemicals would be substituted and life would go on much as before. Global warming, on the other hand, meant fundamental changes to the way people lived. The United States Environmental Protection Agency estimated in 1988 that to cope with climate change new power plants costing $110 billion would be needed over the following 20 years. This was more than the total world CFC bill, just to deal with one global warming problem in one country.

That, in a nutshell, was the problem for American politicians. They had raised the profile of the issue but now that they understood its implications they rather wished they had not. Although the 1988 drought was over, the political interest in global warming was now worldwide and the United States was seen as the great consumer and therefore an easy target. For example, despite the affluence of America and the high standard of living, the price of petrol in the United States remained low compared with the rest of the world. In 1990 the price was one dollar a gallon, half that in West Germany and one third that in France. Japan, which has no oil of its own, charged the equivalent of nearly $4 a gallon. The US was estimated to subsidize its energy industry by $40 billion a year. However, the idea of putting up taxes to reduce emissions was political suicide.

Comparisons with other industrialized countries on carbon dioxide emissions are also instructive. Japan already had very low emissions for an advanced country. Conservation of energy had become a priority in Japan following the 1973 oil crisis because of the country's need to import all its oil. In fact, Japan's example

29

showed what could be done in any developed country. France, with its concentration on nuclear power, also came out well. The figure for the whole of the European Community, at that time ten of the most sophisticated economies outside North America, only produced 14 per cent of the world total compared with 22 per cent for the United States. Whichever way the figures were looked at, each country in the industrial developed world had massively greater emissions than any developing country.

In round figures, it could be said that countries with 20 per cent of the world's population created 80 per cent of the artificial greenhouse gases. The UK is a typical example of this. A relatively small country in terms of area and population, it contributes 21 per cent of Europe's total greenhouse emissions and 3 per cent of the world's. That is the same as the whole of India.

It was from Europe that public alarm about global warming began to focus into political action. Adopting the language of the Montreal Protocol and the CFC negotiations there was talk of targets for reducing carbon dioxide emissions and timetables for doing so. The lead came from the Netherlands and the Scandinavians. The Dutch government was galvanized by its public opinion, naturally alarmed by the prospect of sea level rise, and the Scandinavians were already sold on the idea of targets for other reasons, for example in reducing sulphur emissions to rid them of acid rain.

In the United States, however, the tide had turned against action with remarkable speed. Late in 1988 when concern about global warming was at its height, presidential candidate George Bush had talked about using 'the White House Effect to counter the Greenhouse Effect'. By the middle of the next year the administration had done its sums about the cost of such a venture. It would require the restructuring of the entire energy production sector, hitting core coal and oil industries. It would have repercussions across the world economy. The United States is a country of powerful lobbies and these industries had already mobilized to defend their interests. Their priority was to discourage any moves to control carbon emissions.

The obvious ploy for the new administration was to play for time, in this case demand more research, greater scientific evidence. George Bush several times referred to the need to firm up the science before policy could follow. The green lobby was deeply disturbed by this turnaround.

The scene was therefore set for a row when on 8 May 1989, the Office of Management and Budget confirmed it had altered the Congressional Testimony of NASA's James Hansen. The effect was to weaken his conclusion that enough was known about climate change to justify immediate action. Despite the adverse domestic publicity and international embarrassment, George Bush had clearly decided that he must stall on global warming to protect his political future. Green groups invented a prayer for politicians on the environment – if firm action is needed to save the world please do not require it in my term of office.

Exactly the same kind of process had occurred in Britain. Mrs Thatcher, so recently the champion of the environment, had also been counting the cost to her other cherished policies. Electricity privatization, what she called 'the great car economy', and the removal of regulations in order to open up competition for a free market were the Conservative priorities. All of these clashed with the policies required to combat climate change. The floating of the electricity industry on the stock exchange was particularly important in this period. The legislation was based on the ideology of creating competition, lowering prices and increasing sales. If global warming was to be combated then energy efficiency, and therefore reduced sales, was the most obvious first step that needed to be inserted into the process. This would have required simple redrafting of the legislation. However, a delay and the novel thinking involved would have caused uncertainty in the City, which was used to thinking on traditional lines and gaining a fast buck. In short it would have depressed the share price. The government needed to maximize the money from the sale to finance election-winning tax cuts. This decision to privatize the electricity utilities and subsequently the coal industry was to have far-reaching consequences for Britain's environment policy. Much later, by accident, the Conservatives briefly became heroes, but more of that in later chapters.

In 1989 it was a different story. Against the background of a change of heart by George Bush and Margaret Thatcher the stage was set for a series of international negotiating sessions in which the changing positions of key players became apparent. The United States was cast as the bad guy and Britain was seen as dragging the rest of Europe back.

The first important meeting was at The Hague in March 1989.

It was called by France and the Netherlands and included 12 developed and 12 developing countries. The idea and the result was to call for improved decision-making procedures on international environmental issues. However, its impact was limited because neither the United States nor the Soviet Union were invited.

France, as the current chairman of the group known as G7, the heavyweight industrial democracies, carried the theme forward to the June summit of the organization. The Paris end-of-meeting communiqué, which normally tries to set the tone for the financial and economic outlook for the next 12 months, took the unusual course of devoting one third of its length to the environment. It ranged impressively over a variety of subjects without coming to any firm conclusions.

The United Nations meanwhile was beginning to set up the Conference on Environment and Development for 1992. This had been agreed in principle the previous year as a 20-year follow-up to the Stockholm conference of 1972. As has already been described, one of the results of Stockholm was the setting up of the United Nations Environment Programme. It had presented to the UN alarming evidence on the continuing deterioration of the world's environment. A World Commission on Environment and Development was set up to investigate further and report possible solutions under the chairmanship of Gro Harlem Brundtland, the Norwegian Prime Minister. In 1987 the Commission published its findings. The Brundtland Report, as it became known, introduced the idea of sustainable development, defining a sustainable society as one that manages its economic growth in such a way as to do no irreparable damage to its environment. By balancing economic requirements with ecological concerns, it satisfies the needs of its people without jeopardizing the prospects of future generations. This was a tall order. Organizing the Earth Summit was the UN's attempt to get the message of the Brundtland Report across.

The rows began at once between North and South about the purpose of the conference. Just in case of confusion about these handy jargon labels much loved by journalists and politicians, the 'North' is roughly defined as Europe, North America and Japan, but would include Australia and New Zealand. The 'South' is most of Asia, Africa, Central and South America. The North has an organization, already mentioned, called G7, which discusses and generally agrees broad financial and trade issues. In response the

South set up G77 to represent its views. Originally this did not include China but eventually became a loose bloc of around 120 nations led by China and India which put the South's point of view.

The rows between the two sides in advance of the 1992 conference were about the issues that had first been aired at Stockholm. The Third World priorities were completely different. The word 'development' was added to the name of the conference for that reason, making it the Conference on Environment and Development. The meaning of this rather ponderous title was rather lost on the general public of the industrial world because the press much preferred the shorter and catchier title of the Earth Summit. The nations of the South wanted issues of poverty, aid and transfers of technology to be central to the conference agenda.

Once the conference was arranged, the UN knew that the guarantee of its success lay in its preparation. This time it meant not just a lot of high-sounding words for politicians to say and then sign up to but conventions which committed individual states to real action.

The approach of the Earth Summit provided a deadline to sharpen minds. Work began immediately on a framework climate convention. But what would its objectives be? The first meeting at which the differences in approach between the main players, Europe and the United States, became glaringly apparent was at Noordwijk in Holland in November 1989. This was attended by ministers from 72 countries. The town lies behind a giant sand dune which protects it from the sea. In that setting it was not hard to see why the Dutch hosts were keen to take action to prevent a rise in sea level. The Dutch and the other progressive Europeans demanded that industrialized countries should set themselves a target of stabilizing their carbon dioxide emissions by the year 2000. The Dutch later adopted the target unilaterally as an example to the rest of the world. The UK was the odd man out in the European Community but Japan, the United States and the Soviet Union also resisted targets. Another important argument was over the idea of funding for developing countries to help them combat climate change. The United States, Britain and Japan also opposed this, fearing new aid commitments. The strength of American public opinion on the environment was reflected in the adverse publicity at home to the negative White House stance. President Bush responded by announcing $1 billion extra for climate research and plans to plant a billion trees to take carbon dioxide out of the atmosphere and 'fix' it in the wood. He also announced his intention to host a White House conference on climate change the following spring. This turned out to be a mistake because it again showed American policy to be negative. It also exposed splits in the administration.

The atmosphere in the United States was not helped by an appeal by half the members of the National Academy of Sciences and half the Nobel Prizewinners in the United States asking the administration to act on the threat of global warming. The Environmental Protection Agency and the State Department wanted stronger action but John Sununu, the White House chief of staff, did not. His view was that 'the science of climate prediction was not developed well enough to take actions that might cause economic pain'. He became a pet hate figure for the environment movement.

John Sununu was a powerful figure and his influence pervaded American policy during this vital period. The idea that scientific certainty was required before economic action could follow recurred in the speeches of President Bush. In his major speech to the White House conference, President Bush emphasized the scientific uncertainties, the economic costs, and the need for more research before corrective action. This led to a series of damning press interviews with European government representatives attending who were demanding that the United States set carbon dioxide targets. Particularly prominent was the German environment minister, Klaus Topfer, very popular among environmentalists, who became a dab hand at the impromptu press conference at many future meetings. He had learned early on how to use the media to put pressure on opponents.

The next round of negotiations took place in Bergen, Norway, in May 1990 and marked a high point in the frenzy of name-calling against the United States. The American position had been undermined by the IPCC first working group's interim report, in which 175 scientists from 25 countries agreed that emissions from human activities were substantially increasing the atmospheric concentrations of greenhouse gases. The report predicted an increase in temperature of 1°C (34°F) by 2025 and 3°C (37°F) before the end of the twenty-first century. There were still plenty of uncertainties in the figures but the scientists emphasized the worldwide consensus on the findings.

Bergen took on a new significance. By this time the pattern of

a large attendance of environmental campaigners was well established. The British, until then the European allies of the Americans,
changed their position and accepted the idea of targets. This appears
to be directly as a result of an address by John Houghton, the
chairman of the IPCC working group, to the British cabinet in the
same month. Dr Houghton, who was later knighted, had been head
of the UK Met Office and was a well-respected figure. By his own
account he was listened to in silence by Mrs Thatcher and her 30
cabinet members and officials present. 'Questions and discussion
afterwards demonstrated a large degree of concern for the world's
environment problems', he recalls in his book on global warming.
So for the first time a European target of stabilizing carbon dioxide
emissions was possible. In fact the UK government did not go as
far on targets as other European nations until much later.

The West Germans, under political pressure from the Greens, had
adopted a 25 per cent cut by 2005. The Netherlands announced
it would stabilize emissions by 1995 and cut total greenhouse
emissions by 2000. The British, however, thought this was neither
practical nor possible and so plumped for stabilization by 2005.
This meant, however, by adding together the promises made by
various governments, that it was possible eventually to make a
commitment for the whole of the European Community to return
carbon dioxide emissions to 1990 levels by the year 2000. This
assumed some growth after 1990 and then a reduction as measures
put in place had time to work. This formula for the European
Community ended up being the yardstick against which every other
climate change pledge was made in the run-up to the Earth Summit.

At the Bergen meeting the British escaped most of the flak.
This was partly due to the personality of David Trippier, the then
environment minister. It was hard not to like him and he was
refreshingly straightforward in his views, commenting: 'We could
go for a 2000 target, if we wanted to close down half of the coal
mines in Britain and go for no economic growth.' With this kind of
candour and the change in tack on targets, even if the one chosen
was weak, he earned the British government a breather, the main
focus remaining on the Americans. By this stage, positions were
beginning to harden further. For example, the Americans began to
object to the use of the term 'global warming' and insisted that in
documents it should be referred to as the less contentious 'climate
change'. The words 'targets' and 'timetables' were also to be

avoided at all costs. No commitments were to be made. More science was needed.

Events were moving fast. The same year (August 1990) the First Assessment Report of the IPCC was approved in Sundsvall, Sweden, in the run-up to the Second World Climate Conference in Geneva in November. The Sundsvall meeting was already as much about politics as science. The IPCC's third working group, which was to explore responses to climate change, was criticized for lacking bite. It was said openly that the Americans had used their chairmanship of the group to make sure that measures that would have far-reaching economic consequences were kept off the agenda. Lobbying outside the meeting from industry and environment groups was intense. It was now the habit for sessions to go late into the night. Sentences, or even single words, which were not agreed would be put in square brackets in the provisional texts. The United States managed to have the words 'global warming' expunged. Not that the Americans were alone in wanting to water down the findings. The oil-rich states of the Middle East, which later emerged as a formidable bloc, were also keen to avoid any action that might hurt their staple export. At the time they were content to leave the Americans to take the lead in resisting action and so escape the odium that went with that position.

The Russians, who because of the turbulent situation in the Soviet Union were not really significant in this political struggle, did make a strong scientific input, however, and had done so throughout the cold war, their experts sharing all available information. Nevertheless, there was one occasion when politics nearly interfered in this process. In early negotiations the Russians refused to accept that rapid global warming was a bad idea. It emerged that in Moscow it was felt that it might be a good thing for Communism. If global warming meant that the Great Plains of the United States were too hot and dry to grow wheat then Russia could replace its political rivals as the bread basket of the world. After all, if the Russian steppes got warmer they might become the new prairies. Fortunately, in this case the scientists were able to persuade the politicians in the Kremlin that this was not sound reasoning and Russia rejoined the consensus that warming should be slowed down as much as possible.

The World Climate Conference in Geneva in November 1990 was a political jamboree, attended by a throng of concerned world leaders, including Margaret Thatcher and the king of Jordan, who rightly had become concerned about the threat to his country's already sparse water supply from the effects of climate change. It was an opportunity to be seen as a world statesman and many politicians attended. The tone was upbeat; everyone was agreed that it was a problem that must be tackled and would be. Mrs Thatcher, using her knack for political gestures, became the first head of state to pledge her attendance at the Earth Summit, and was followed by many others in the months to come. The conference urged countries, in advance of the Earth Summit, to develop programmes, strategies and/or targets for a staged approach for achieving reductions of all greenhouse gas emissions over the next two decades and beyond. It was a pious hope and perhaps the high spot of political rhetoric on the subject before the hard work of trying to turn it into real agreements with teeth took over.

The road to Rio and Earth Summit was to prove long and hard, however. The scientists continued to beaver away. Thanks to President Bush and Mrs Thatcher research in both the United States and Britain was given plenty of cash – in both cases to buy time politically to avoid action to cut greenhouse gas emissions. The IPCC were to produce a science update and confirmation of their earlier findings in time for the Earth Summit.

On the political front, opinions began to harden. The Arab nations had appeared at Geneva and made it clear that they were opposed to the preparation of a climate change convention, at least if it had any teeth to limit carbon dioxide emissions. They became the close allies of the United States when the Intergovernmental Negotiating Committee (INC) began a series of meetings to try to thrash out a convention. In 1991 the committee met no less than four times, once in Washington and Nairobi and twice in Geneva. The same issues came up again and again. Just to get over their point 40 developing countries issued the Beijing Declaration, in which their demand for resource transfers to deal with global warming was reinforced.

President Bush continued to be the stumbling block to any mention of targets and timetables. He had been frightened by a Council of Economic Advisors report in 1990 which estimated the cost to the United States of cutting carbon emissions by 20 per cent by the year 2100 to be between $800 billion and $3.6 trillion. The figures were frightening if meaningless in their magnitude, but the message was a repeat of John Sununu's statement: 'High priority in the near future should be to improve understanding in order to build a foundation for sound policy decisions. Until such a foundation is in place, there is no justification for imposing major costs on the economy in order to slow the growth of greenhouse gas emissions.' This stiffened President Bush's resolve and the United States refused to budge. In the last six months before the Earth Summit the United States made it clear that the President would not attend if the proposed Climate Change Convention bound its signatories to specific obligations. In the end the Americans won and the convention was far weaker than the rhetoric of 1990 had suggested. A Climate Change Convention at last existed, however, and it became a key part in giving the Earth Summit some meaning, beyond its subtitle as the biggest conference the world had even seen, a title taken over from its predecessor in Stockholm 20 years before.

# 4

---

# The Earth Summit
# and the signing of the
# Climate Change Convention

THERE WERE 118 HEADS OF STATE gathered at Rio de Janeiro in June 1992. In the conference corridors it was possible to rub shoulders with kings, presidents, prime ministers, heads of religious orders and celebrities. Leaders like George Bush and Fidel Castro, who would not normally share any political platform, were signing up to the same pledges to save the earth.

Altogether 178 countries attended what was properly titled the United Nations Conference on Environment and Development (UNCED). They sent a total of 7000 registered delegates, along with 1300 representatives of non-governmental organizations whose intention was to bend the delegates to their point of view. During the 12 days of the conference 8000 journalists filed reports back to their home countries. They universally described UNCED as 'the Earth Summit'. A separate 'alternative' summit called the Global Forum was held on the seashore at Rio, 50 km (30 miles) from the main conference at the Rio Centro. It was attended by 17,000 foreign participants. Every shade of opinion was represented. There were native Indian tribes, religions which thought the world was about to end, and a vast array of environment groups who tried to draw up an alternative summit document but found they could not agree amongst themselves.

The whole thing was an extraordinary event, brilliantly organized by the Brazilians. The fact that it functioned relatively smoothly with virtually no security problems was remarkable in itself. There were tanks on the streets to prevent trouble. The prostitutes, muggers and street children had been cleared away from Rio's famous beaches so that the delegates were not troubled, and the journalists staying in the hotels were not tempted to write about them. The crime rate for the two weeks of the conference plummeted because of the massive presence of soldiers but the starving street children still begged for leftovers at the beach restaurants and wolfed them down before the police could catch them. It was a useful education in itself for first-time visitors to the developing world.

But the big question, at the end of all the razzmatazz, was what did the Earth Summit actually achieve? Even at this distance, four years later, no one is quite sure. There were thousands of pious words spoken, and often obscurely worded documents agreed. From the point of view of this book the UN Framework Convention on Climate Change was the most important of these and, compared with most of the others, it was a model of clarity. A second useful convention on preserving biodiversity was also agreed and signed. This was also worth celebrating because it tackled another fundamental and difficult problem. Again it was completed in the teeth of opposition from the Bush administration. Some people have already gone so far as to say that the whole summit was worth organizing in order to achieve a climate convention. Certainly it was a milestone in the battle against global warming and in 20 years hence I expect UNCED will be celebrated for that and the Convention on Biological Diversity if nothing else.

But UNCED was about much more and most of the discussions touched subjects which have a direct bearing on climate change. For many people present the conference was an education in hearing other people's points of view. The Brazilian President Fernando Collor de Mello put it this way: 'You cannot have an environmentally healthy planet in a world that is socially unjust.'

The most obvious area which affects climate and caused the most fundamental of disagreements was forests, closely followed by deserts. The row over a proposed forest convention was a classic North–South confrontation. There was a strong feeling in the industrialized countries of Europe and North America that developing countries like Malaysia and Brazil must be stopped from destroying their tropical forests. Malaysia's Prime Minister Mahathir Mohamad took a hard line. If the industrialized nations thought rainforests were so important for biodiversity and carbon dioxide storage why did not the rich, carbon dioxide producing countries pay for the service of preserving those forests, instead of bullying poor countries not to utilize one of their few natural resources. He then went on the attack, pointing out that Europe was once covered in forests until the local residents had cut them down.

In the end, a document of 'forest principles' was agreed which amounted to another set of good intentions, but with the hope of organizing a convention later. This has still to materialize.

Deserts were not on the North's agenda at all until African states insisted that this was their priority. Work on a desert convention was promised and indeed progress has been made in this area, both in research to find out what is happening in certain areas and in planting and management schemes to see if the spread can be halted or reversed.

Both these issues will be important in future climate negotiations. It would be surprising if there is not some linkage between the progress on the Climate Convention and the need for full-blown conventions on these two subjects. As we have already seen with President Bush's as yet unfulfilled pledge for a billion new trees, forests have a vital role in fixing carbon dioxide from the atmosphere back into living things. Scientists call them 'carbon sinks'. Deserts, on the other hand, are directly linked to the process of climate change, according to the IPCC. Although all sorts of other activities, including poor agricultural practice, cause deserts, the effect on rainfall and evaporation caused by global warming is going to be significant if it is not already.

The Biological Diversity Convention does not have such direct connections to climate change except that one of the greatest threats to any species must be a change of climate that occurs too fast for it to adapt. In other words, to protect the diverse habitats on which biodiversity depends, climate change needs to be addressed as a priority. Even without climate change the extinction rate is frightening. In the 1970s it was estimated that 30,000 species a year were being lost. In the 1980s it had accelerated to 50,000 species a year. At the Summit it was estimated that by 2020 between 10 and 20 per cent of the earth's 10 million species of plants and animals will have disappeared for ever. That makes a mockery of the idea of sustainable development.

Apart from these three specific areas linked with climate change – forests, deserts and biodiversity – there was a mammoth document called Agenda 21 which, if fully implemented, would solve many of the world's inequities and its problems. The document was described at the time as a blueprint for environmental action into the twenty-first century. It was full of hopes and dreams, but sadly short on specifics, for example, what should be done about population growth. There was a great deal of argument over birth control with the Vatican, which had the support of some Catholic developing countries. As a result an issue which was arguably central to the whole debate was mentioned only as requiring 'appropriate demographic policies'. That was the most glaring example, but wherever there was a strong objection from one vested interest or other to clear wording and targets there was an appropriate fudge. In the end the object was to produce a document that everyone could sign even if parts of it were rendered almost meaningless in the process.

Nevertheless, it would be wrong to be too cynical about this grand event. As at Stockholm, only more so in this case, the whole world focused on the Summit and in doing so educated vast numbers of people about the environment. It also created a number of new bodies and systems for tackling some global problems. These have since cranked into action, some with good effect. Unfortunately for the environment, the timing was bad on the economic front. The boom of the late 1980s when the idea of the conference was first conceived had turned into a recession across the industrialized world. With economies on the downturn the issue of aid for developing countries, and particularly extra aid, was a thorny one. Initial hopes of a big injection of new cash for environmental projects from the industrialized world were dashed. There was much discussion of an existing United Nations target of 0.7 per cent of gross national product to be given in aid by developed to developing countries. Only Sweden exceeded it, and most affluent nations were well below it. In fact the British overseas aid budget had been steadily shrinking under Mrs Thatcher's premiership and went down further under John Major. Attempts to insert a target for industrialized nations to get up to the 0.7 per cent by the turn of the century were firmly rejected and it became another vague 'as soon as possible'.

President Bush, who finally agreed to attend, probably wished he had stayed at home. As has already been described, the Americans had made themselves unpopular by refusing to agree to targets on global warming. The conference had begun with headlines across the world about the decision of the United States not to sign the

Convention on Biological Diversity. The reason was the cost in aid to developing countries to help them save their internationally important habitats. A memo revealing sharp divisions in the United States administration over the funding for biodiversity was leaked at the conference. It was from William Reilly, well-respected head of the Environment Protection Agency in the US and its head of delegation at the conference. Reilly was clearly uncomfortable defending his government at the daily press conferences. In a another memo (leaked later) he described his experience of being President Bush's front man at Rio as 'like taking a bungee jump while somebody cut the line'.

Another miscalculation was the announcement before President Bush's arrival of $150 million dollars in aid for forests. This was seen as an attempt to divert attention from the administration's failures in other areas. Each head of state was given ten minutes to address the world and the President spent his allotted time defiantly defending America's record and saying at one point 'I did not come here to apologize'. It is worth noting that the US Clean Air Act of 1990, which was to reduce sulphur dioxide emissions by 35 per cent, also had the effect of cutting back on greenhouse gases. It was claimed at Rio by the American press relations team that it would probably mean that the United States would peg carbon dioxide emissions at 1990 levels by the year 2000. This has proved a false hope. In the end the United States had a miserable UNCED. They were able to sign the Climate Change Convention along with 153 other states plus the EC, but had to take the blame for watering it down. Notable absentees from the signing ceremony were the big oil producers including Saudi Arabia, Kuwait and the United Arab Emirates. The Convention on Biological Diversity attracted 155 countries, the much noted missing signature here being that of President Bush.

But after all the hullabaloo of the Earth Summit, what did the Convention on Climate Change actually say? All the attention during negotiations had been on whether industrialized countries would accept targets for reductions of carbon dioxide and time-tables to achieve them. Particularly whether the United States would play ball with this idea. But putting that issue on one side, re-reading the Convention gives new heart. Considering the difficulties in the drafting process and the many conflicting national interests involved, common sense shines through. Following the lead taken by the Montreal Protocol on ozone, the negotiators have built in a 'precautionary principle'. This is a phrase that has since become so overused that people do not think what it really means. In this case it means that people who sign the convention will not wait to have proof that the world is being seriously damaged by climate change before taking action. It makes another point. There are a number of actions we should be taking for the good of the world which also incidentally mitigate climate change. Examples of this are planting more trees and large-scale energy efficiency projects. In other words, we do not need and cannot afford to wait for confirmation before we act.

By international treaty standards the stated objective of the Convention is brief, to the point, and covers the difficult ground of continued scientific uncertainty with great skill. The Conference of the Parties referred to in the text is the future meetings of countries that have ratified the Convention whose job it is to implement it. Here is Article 2 in full.

> The ultimate objective of this Convention and any related legal instruments that the Conference of the Parties may adopt is to achieve, in accordance with the relevant provisions of the Convention, stabilization of greenhouse gas concentrations in the atmosphere at a level that would prevent dangerous anthropogenic interference with the climate system. Such a level should be achieved within a time-frame sufficient to allow ecosystems to adapt naturally to climate change, to ensure that food production is not threatened and to enable economic development to proceed in a sustainable manner.

This adds up to a pretty tall order. Four years later progress has been so slow it seems unlikely that the countries who signed up to the Convention will achieve its objective, but Article 2 leaves them no doubt what they are aiming at. The Convention also tackles the thorny problem of unfairness between rich and poor countries. It acknowledges that the rich countries created their wealth in part by pushing into the air vast amounts of greenhouse gases long before the consequences were understood. The United States tried to avoid being held responsible for this by saying that present populations cannot be forced to pay for the mistakes of their fathers, but they were overruled.

Developing countries feared being told that they should curtail their own industrial development because the atmosphere's safety margin was already used up. The Convention responds by noting the responsibility of the developed world for the problem and saying: 'Accordingly, the developed country parties should take the lead in combating climate change and the adverse effects thereof.'

At the end of the Convention there are two lists, or Annexes, containing the names of the countries who have to take this 'lead'. The first annexe contains those whose contribution to producing the greenhouse gases means that they have to adopt measures immediately to mitigate climate change by reducing emissions. These are the countries responsible for the current excess of carbon dioxide in the atmosphere since they became industrialized first. These so-called 'advanced' economies, it was hoped, would be subject to specific binding targets to reduce their emissions. Instead there is a general 'aim' of returning emissions to 1990 levels by the year 2000. This annexe includes countries of the old Eastern Soviet bloc – what by then were called 'countries that are undergoing the process of transition to a market economy'.

Annexe II is a list of many of the same countries but these exclude former Communist states and all are members of the Organization of Economic Cooperation and Development (OECD). They agreed to support climate change activities in developing countries by providing financial support above and beyond any they were already providing to these countries for other reasons.

The Convention notes that the poorer nations had the right to economic development and that would mean more greenhouse gases. It also makes special provision for aid to those countries particularly vulnerable to sea level rise, or with mountain or other ecosystems that would be damaged by climate change. The convention even recognized the states which depend on income from coal and oil which would 'face difficulties' if energy demand changes. This hand of friendship was extended despite the fact that these states had done their best to weaken the Convention and at the time of the Earth Summit some of them did not see fit to sign it.

Nevertheless, the Third World and countries of the South are not let off the hook entirely. It is perfectly clear that tackling climate change without ultimately controlling their emissions as well is a waste of time. The Convention therefore takes the first and comparatively painless step of asking them to calculate how

## Carbon emissions

World carbon emissions from fossil fuel burning, 1950-94, million tonnes

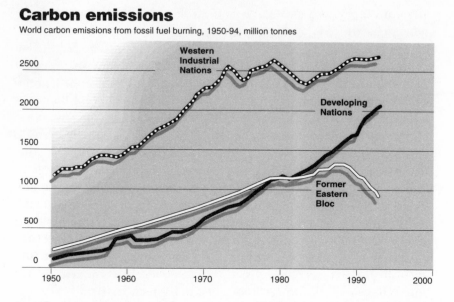

much greenhouse gas they currently pump into the atmosphere. This may sound a relatively simple process but many countries had no expertise in how to do it.

Sophisticated Western economies already know precisely how much oil is imported and exported, how much is stored and how much burned. This is nothing to do with climate change, it is to do with keeping tabs on whether everyone is paying the correct tax. Simple sums can be done about how much carbon dioxide will be produced by a given amount of coal being burned in a power station. In many less-developed countries these kinds of figures are not easily available. They also have different problems. For example, in rice-growing areas the methane produced from paddy fields may make a much higher contribution to total greenhouse gases than burning fossil fuels. It is also a much more difficult figure to calculate.

One of the bits of useful and comparatively inexpensive pieces of aid made available under the Convention was the expertise of creating national inventories of greenhouse gases. Every country that ratified the Convention had a duty to produce these within six months, and although many were not produced on time and some were far more thorough and believable than others, it at least gives some idea where the world is going.

The words 'signing' and 'ratifying' have crept into the above discussion as if they were interchangeable. In fact, the adoption of any convention is a two-stage process. After it has been signed the next step, before it can be implemented, is for countries to get their parliament or legislature to adopt the convention into domestic law. In this case it was stipulated that the Climate Change Convention would come into force on the 90th day after the deposit of 50 instruments of ratification by signatory countries. Sometimes it takes years for enough countries to ratify a convention to allow it to pass into international law and in the meantime virtually nothing happens. This case was different. There was an assumption that the process would continue fairly rapidly. In addition, there was so little time between 1992 and the year 2000 to tackle the first phase of the problem that most advanced countries were already working on their greenhouse gas emissions. If they were to stand a chance of getting them down again to 1990 levels by the year 2000 they needed to start work the moment the Earth Summit ended. In any event their local and vocal green groups would make sure that they adhered to the spirit of the Convention they had just signed.

The optimists proved right and the ratification process for both climate and biodiversity conventions was remarkably rapid. The Convention on Climate Change took effect on 21 March 1994, less than two years from when it was opened for signature. By the time the first Conference of the Parties (COP1) had been organized a year later in Berlin, 165 states had signed and 120 ratified.

Thus had begun the real testing time for the Convention. As has been discussed, there are no binding legal requirements to peg or cut greenhouse gas emissions in the document. What had been put in place was an in-built ratchet effect. We have already discussed the national plans that had to be submitted giving a snapshot of the greenhouse gas emissions and a rundown of government plans to mitigate them. What the Convention also stipulated was an annual meeting of the parties to discuss further and better action. The single most glaring omission was the lack of any kind of reference to what would happen after the year 2000. This was clearly going to be the major preoccupation of the parties to the Convention.

But aside from this headline-grabbing problem, machinery also had to be developed for bringing the developing world more actively into the process. It was clear from the beginning that unless the industrialized world took its obligations seriously, the developing world would not be interested in coming to the party. The stark reality that faced COP1 in Berlin in 1995 was that doing something about climate change had not got any easier. It was going to be just as expensive and politically painful for the countries of the North to do something positive about reducing greenhouse gas emissions as it had been in 1990 or 1992. But just as important was that the situation was getting more urgent. The science was firming up all the time; global warming was very bad news. The task then that faced COP1 was to move towards a protocol. This is an addition to the treaty that details what will be done about climate change by the contracting parties to the Convention after the year 2000. It must begin to tackle global warming seriously.

As will be seen from the next two sections, the scientists have continued to do their job, and we will look at the latest research completed on how things are heating up and the oceans rising. Then we will discuss the latest on present and likely future effects. The scientists are continuing the research, but they are pointing out that, having provided some of the answers, it is now up to the politicians to take the process forward.

How the politicians came to grips with that challenge in Berlin and the arguments raised there and since are in the final section. Examined too are the performances of some key governments and sectional interests. What are major players in the oil industry up to and will the insurance world enter the fray and alter the political balance? What happens next? If the politicians cannot get their act together and tackle climate change, will the new millennium begin with the future of civilization in doubt? Some of the answers are in the next three sections.

# Part Two

---

# THE SCIENCE

# 5

---

# The greenhouse effect
# keeps us alive

ANYONE WHO HAS SAT on a warm rock as the sun goes down knows that the warmth absorbed during the day is slowly released back into the air. On a clear day the heat from the sun, in the form of short-wave radiation, slices right through the atmosphere and heats the earth and oceans. The tropical regions get hottest because the sun is directly overhead. The mid-latitude and polar regions gain some direct heat and then get extra benefit because of ocean currents and winds which circulate some of the tropical warmth to colder parts. In a climate in perfect balance the amount of heat released back into space equals the amount gained from the sun. However, the system is far from simple, as the constant variation in temperature and weather from one day to the next tells us. This is why there is so much argument about exactly how much our interference will cause the climate to heat up. One complication, for example, is that the hotter something is the quicker it cools, so exactly how, and how much, extra heat escapes back into space, are also matters for discussion.

One simple fact makes global warming possible. The heat that is radiated back from the rocks and oceans is at a different wavelength from the incoming sunlight. This returning heat is known as long-wave radiation and has different properties from the incoming short-wave radiation. This difference accounts for the natural greenhouse effect which makes life on earth as we know it possible in the first place. The long-wave radiation does not travel back out through the atmosphere like the sunlight coming in because it is partly absorbed by some naturally occurring gases in the atmosphere. The main one is water vapour, mostly in the familiar form of clouds. Then there is carbon dioxide, methane, nitrous oxide and ozone. These gases are more efficient at absorbing outgoing long-wave radiation than incoming short-wave radiation. They therefore trap the heat; hence the natural greenhouse effect. The heat is then distributed round the globe by direct radiation, as from the rocks, air currents, evaporation, cloud formation and rainfall.

Thus the greenhouse effect caused by the difference in absorption rates of long- and short-wave radiation makes the earth far hotter than it would otherwise be. Estimates vary, but the general consensus is that the earth is 33°C (91°F) warmer than it would be without the natural greenhouse effect, that is with the mixture of gases in the atmosphere which existed before the industrial revolution began in about 1750. In any event, the world would be quite uninhabitable by its current mix of animal and plant life if it were not for this natural global warming effect. It gives us a current average global surface temperature of 15°C (59°F).

If everything else remains constant, an increase in any one of the greenhouse gases would cause more heat to be trapped. For example, if we do what we have been doing since the beginning of the industrial revolution, increasing the carbon dioxide in the atmosphere, logic tells us that the world will become warmer.

A doubling of the concentration of long-lived greenhouse gases in the air, which is projected to occur early in the next century, would, if nothing else changed, reduce the rate at which the planet can shed energy to space by about 2 per cent. Because this would not affect the rate at which energy from the sun is absorbed, the existing balance would be upset. The 2 per cent may not sound like much, but energy from the sun arrives in vast quantities. Over the entire earth it would amount to trapping the energy of some 3 million tonnes of oil every minute.

Similar greenhouse effects also occur on our nearest planetary neighbours, Mars and Venus. Mars is smaller than the Earth and possesses by comparison a very thin atmosphere. The pressure at the surface of Mars is about 1 per cent of that on Earth. However, the Martian atmosphere consists almost entirely of carbon dioxide, which through its own greenhouse effect warms the surface by about 10°C (50°F). Venus has a much thicker atmosphere, also largely

# The greenhouse effect

There is a natural 'greenhouse effect' that permits life. The greenhouse gases such as carbon dioxide, chlorofluorocarbons and methane disturb the natural balance of the atmosphere by trapping additional heat radiating from the earth's surface.

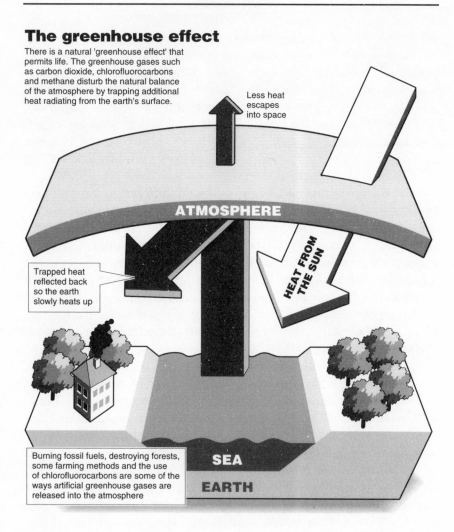

Less heat escapes into space

ATMOSPHERE

HEAT FROM THE SUN

Trapped heat reflected back so the earth slowly heats up

Burning fossil fuels, destroying forests, some farming methods and the use of chlorofluorocarbons are some of the ways artificial greenhouse gases are released into the atmosphere

SEA

EARTH

composed of carbon dioxide, with a surface pressure 100 times that of the Earth. The resulting greenhouse effect is very large and leads to a surface temperature of about 500°C (932°F) more than it would otherwise be.

Scientists first became interested in the idea of fluctuations in the quantity of greenhouse gases on earth in an attempt to explain previous climate changes. For example, during the ice ages, where much of Europe, as far as southern England, was covered in ice sheets, presumably the quantity of greenhouse gases had gone down. At other times we had more carbon dioxide in the atmosphere. It was, for example, far warmer when the dinosaurs roamed the earth. One theory for their apparent sudden disappearance was that a giant asteroid hit the earth, throwing huge amounts of dust into the atmosphere and blotting out the warming rays of the sun. This lowered the temperature of the earth so much that the dinosaurs died out. Or was it simply a reduction in the greenhouse gases over a few centuries that made the earth go colder and gradually killed them off? Scientists think it was far more complicated than that. For example the earth's orbit in relation to the sun and its tilt on its axis also alter the climate. Nevertheless, if the amounts of greenhouse gases in the air are mapped out for the past 160,000 years including the last ice age there does appear to be a direct relationship to the temperature of the earth. As the carbon dioxide goes up and down so does the temperature of the earth's surface.

In the remote past human beings are said to have evolved when the world temperatures dropped. About six million years ago this caused rainfall to decline and ape-like hunters in the Great Rift Valley of Africa used to sheltering in trees were forced out on to the grassland. The apes found themselves on an empty plain, much colder and drier than they were used to, and vulnerable to predators. According to the theory they had to either adapt or face extinction, and in response humans made two evolutionary jumps. The first was to walk upright, allowing the forearms to be used for carrying children and food. Secondly, the size of the brain increased, promoting the use of tools, and the facility to think through problems rather than rely on instinct.

Shifts in climate have caused many problems and bonuses since. For example ice ages which shrunk the sea by piling up water at the poles enabled humans to cross into new continents by land bridges. The same sort of changes also brought disaster. For example in recorded history the Little Ice Age in the Middle Ages brought famines, uprisings and withdrawal of northern colonies in Iceland and Greenland. It is the success as a species in dealing with these problems which has led to the current huge numbers of the human race. These days large-scale migration is no longer an option without serious political consequences. Yet it is our success, our civilization, our industry, electric light and transport, that has created our current problem.

At least we now know what we are doing to increase carbon

dioxide levels in the atmosphere, but what caused them to rise and fall in the recent geological past is still not properly understood. Huge and complex chemical reactions are taking place 24 hours a day all over the globe in which the gases in the atmosphere are being changed and replaced constantly. Living creatures in the oceans, forests, prairies and tundra are all using the atmosphere and its gases to survive. If it was all in perfect balance nothing would ever change, but it is changing and probably always has been.

It is believed that carbon dioxide once made up the majority of the atmosphere. The mixture of the air, if one could call it that, was formed as a result of the gases spewed out of the constantly erupting surface of the planet which was a mass of active volcanoes. Because so much of the sun's heat was trapped by these gases the earth was a very warm place. It began to change when plants, or tiny specks of living things, used the carbon dioxide and the energy in sunlight together with the minerals and the nitrogen to grow. They discharged oxygen back into the air as a waste product. Over billions of years they evolved and grew into larger plants. All the time they were taking carbon dioxide out of the atmosphere, turning it into carbon and using it to build leaves and stems. Thus they fixed the carbon in masses of vegetation, that in turn became vast jungles that sunk back into the swamps. As one generation of plants grew on top of its decaying ancestors a layer of carbon was formed. This process still continues and it becomes recognizable to us as a layer of fibrous decayed matter, a black carbon-laden substance called peat. If left undisturbed the process continues and through millions of years of geological time the peat becomes compressed and turns into oil and coal. Oil and coal are familiar substances, and form an important part of this story because we are now burning them and turning them back into carbon dioxide.

But that is only part of the equation. Carbon dioxide is taken out of the atmosphere by photosynthesis all the time. When plants decay, or in the case of some forests are chopped down and burned, or are eaten by animals, the gas is released again. The same process happens at sea. Billions of tiny sea creatures called plankton are also using carbon dioxide to grow. All the time they are living, being eaten and dying in the two thirds of the earth's surface covered by the oceans. Precisely how much carbon is gained or lost in this way is still not known, but it is known that some carbon is being taken out of the system when these creatures die. Their bodies sink to the bottom of the sea where some of the carbon remains in the mud. There is also the straight absorption of carbon dioxide into seawater. Most estimates say that about one third of the carbon dioxide being released into the atmosphere at the moment is absorbed by the oceans.

This process of using and then fixing carbon dioxide into carbon through billions of years has changed the composition of the atmosphere and the face of the planet. If the atmosphere had not changed because of plants and had remained heavily laced with carbon dioxide, it is estimated that the world would be 45°C (113°F) hotter than it is now.

As life evolved, the amount of carbon dioxide in the atmosphere reduced to a few parts per million and the oxygen increased until a new niche appeared for other creatures to exploit. Animals, eventually including humans, used the oxygen to power their existence and in turn breathed out carbon dioxide. Once a balance was struck then the climate of the world reached roughly what it has been in the last few hundred centuries. By the time humans had begun to walk upright, nitrogen was by far the dominant gas. Carbon dioxide had shrunk to a tiny proportion of the atmosphere, and the oxygen had grown to about 21 per cent.

In fact, carbon dioxide had shrunk to roughly 0.03 per cent of the air. In the last couple of thousand years it appears to have been stable at 280 parts per million by volume. That suited us and the rest of the plants and animals fine. It made the world warmer than it would be without the greenhouse effect, but not too hot. The problem is that we seem to have upset the balance. In the last 200 years, because of human activities, the proportion of carbon dioxide has risen from 280 parts per million to 360 parts per million at the end of 1995. Thus, in the time since the beginning of the industrial revolution, when we started to burn coal rather than wood to fuel civilization, the proportion of carbon dioxide has risen by 27 per cent. As industrial activity increases, particularly in the developing world, the rate of carbon dioxide release grows too. Currently it is increasing by about 3 parts per million a year.

The other gases that fuel the natural greenhouse effect have also increased. Methane is a much rarer gas, measured in parts per billion in the atmosphere rather than parts per million, but it has a much greater warming effect – 21 times that of carbon dioxide by volume. While carbon dioxide has increased by 27 per cent in the past 200 years, methane has increased 145 per cent – from 700 parts per billion to 1720 parts per billion. It is calculated to be responsible for about 15–20 per cent of the enhanced greenhouse effect.

Despite the steepness of the rate of increase of methane and its greater potential for global warming, molecule for molecule, scientists seem less bothered about this rise than that of carbon dioxide. The reason is that carbon dioxide survives longer in the atmosphere than methane, between 50 and 200 years. That means that once the increase in volume of gas has occurred, it is much harder to reverse the process. Methane, on the other hand, is destroyed by chemical processes within 12–17 years, so the damage is potentially more easily reversible. Another reason that methane rates much less of a mention is that the causes of its production are more complex, and often difficult to control even if there was the political will to do so. Sometimes methane can be downright embarrassing. Cattle produce large quantities of methane via their digestive processes by belching. Termites made the headlines in March 1995 when it was disclosed that their flatulence may be responsible for up to a fifth of the world's production of methane. It is the sort of story journalists love. Dr Paul Eggleton, from the Natural History Museum in London, was reported to be bringing tropical termite farts back through customs from the Cameroon in West Africa. They were in jars marked 'forest air'.

Dr Eggleton and his colleagues found that the 3500 species of termites in the tropics produce methane as a result of specialist bacteria in their guts. Termites are thought to make up 10 per cent of the volume of all animals in the tropics. They feed on wood, dry grass, decaying leaves, animal dung and humus. The ones that live in the tropical forests emit more methane than those living in deforested areas. Chopping down the tropical forests to kill off the termites is not the answer, however. The loss of the methane-producing termites would be far outweighed by the carbon dioxide released by burning trees. The scientists calculated that the termites produce between 20 and 80 million tonnes of methane a year (note the huge uncertainties) but say this is part of an ancient natural process and should not be controlled.

The amount of atmospheric methane has only started to rise since the industrial revolution, and the increase in quantity follows very closely the growth of human population since that time. If no attempt is made to control human-related sources it is reasonable to assume that this trend will continue. The United Nations estimate that the world's population will double in the next century, so unless action is taken enhanced warming from methane will double over that time.

Human activity releases methane trapped in the ground via coal mining; the gas is also mixed in oil and is burned more and more widely as a constituent of natural gas. This brings its own problems since transporting natural gas in pipelines naturally leads to leaks and an increase in atmospheric concentrations. The termite output, at about one fifth of the world's total, is still less than that produced from cattle and sheep, about 100 million tonnes.

Methane emissions have also been increasing for other reasons. Particularly large quantities come from rice paddies and scientists are trying to find new growing techniques to reduce the production of methane in agriculture. Another source of methane is rotting rubbish in landfills. This problem is being tackled in some indus-trialized countries by capping modern landfill sites and turning the captured methane into a fuel for generating electricity. This reduces both the methane release and the need to burn fossil fuels. The methane released from coal mining is significant and there is a constant battle to control it.

All the artificial emissions can be reduced but there is another, possibly uncontrollable feature. There are millions of tonnes of methane locked in the permafrost of northern regions, in what are called gas hydrates. These are ice-like mixtures of water and organic gases which only form at high pressure and low temperatures – in this case probably during the last ice age. But because the world is warming the permafrost is beginning to melt. Researchers who have dug these hydrates out of the ice and let them warm up have discovered that a single cubic metre (1.3 yd$^3$) of methane hydrate can produce 160 m$^3$ (210 yd$^3$) of gas. The release of methane in large quantities from the arctic tundra could make a substantial difference to global warming, in what some people call a 'runaway greenhouse effect'. Research is continuing into this alarming possibility.

Nitrous oxide, known as laughing gas and used as an anaesthetic, is another greenhouse gas. Its role is minor but it has a relatively long atmospheric life of 120 years and is steadily growing in volume at about 0.25 per cent a year. Its concentration has risen from 275 to 310 parts per billion by volume in the air, about 13 per cent, since pre-industrial times. Again the exact reasons for this rise are not clear. The burning of wood and other fuels is partly to blame, plus the production of the gas in the ground following the application of ammonia-based fertilizers. Various industrial processes, including the production of acid, also give off nitrous oxide. Artificial sources would have to be reduced by 50 per cent according to the IPCC if levels are to be stabilized. This is not yet seen as a key problem but it is an area where information is inadequate for politicians to make any kind of effective decision on reduction methods. Only time will tell whether this gas will become more prominent in the discussions to come.

The above gases are those that occur naturally and account for the way the world has developed without our help. All we have done is added to them and so enhanced a natural process. Next we look at the contribution made by CFCs, an entirely artificial phenomenon. These gases were good servants but appear to have turned into monsters.

CFCs or, to give them their full, ghastly title, chlorofluoro-carbons, did not exist at all until they were invented in the 1930s, as ever with the best of intentions. For years they were seen only as a great bonus to industry but they turned out to be creating a potential disaster too. Scientists and industry have since done a thorough job on both understanding the causes for this concern and substituting other inventions for this batch of gases.

They were created as extremely stable, non-toxic and non-inflammable gases which were easy to liquefy under pressure. This meant that they were particularly suitable for use in refrigerators and air-conditioning and as a propellant in aerosol spray cans. In fact they were a great boon, allowing safe domestic refrigerators to be readily available for the first time. A number of different types were developed for different purposes and given different numbers, CFC 11 and 12 for example. Other uses included blowing agents for foam rubber and rigid polyurethane foam in furniture.

CFCs were designed to make our life simpler but in fact have made it much more complex, and have left us with serious dangers to health and world food supplies. CFCs have been recognized as responsible for both severe damage to the ozone layer and potentially exacerbating global warming. They were first identified as harmful to the ozone layer in 1974 but at that stage it was all just theoretical. The United States banned them in aerosols as a precaution in 1978 but it was not until 1985 when a 'hole' in the ozone layer was mapped in the Antarctic that international action began. It is still hard to credit that the contents of an aerosol or the back of a domestic fridge can be so harmless at ground level and yet can drift up into the upper atmosphere and, through a complex chemical reaction caused by sunlight, damage the ozone shield which protects us from the ultraviolet light of the sun.

The politicians met and were galvanized by the science. It was clear that ozone depletion was a real threat. Since then, measurements at both poles show that the protective ozone is getting thinner each year, and there is a subsequent thinning of the protection across the globe. As related earlier, governments reacted surprisingly quickly, worked out what to do and had signed the Montreal Protocol by 1987. Subsequently more ozone depleters were identified, the science refined and target dates for various chemicals to be reduced or substituted agreed.

But while it was the effect that CFCs had on the ozone layer that grabbed the immediate attention of scientists and politicians alike, it was also recognized that CFCs are global warming gases themselves. The prime worry was that the chlorine contained in CFCs, once in the upper atmosphere, acts to destroy some of the ozone. As discussed earlier, the ozone layer is a vital part of our planet's life support system in its proper place in the upper atmosphere. Here it acts as a shield for the earth against some of the more harmful ultraviolet light from the sun. At the same time ozone is a natural greenhouse gas which prevents some heat escaping back into space. A sort of upper blanket.

At ground level ozone assumes a different role and is regarded as an unwanted pollutant because it damages human lungs and is toxic to plants. Here, where it is not wanted, quantities are rising. The gases responsible for creating low-level ozone are carbon monoxide, the oxides of nitrogen (all found in car exhaust fumes), methane and other hydrocarbons. The added ingredient is sunlight, which is why the smoke and fumes humans emit in going about their daily lives come back to haunt us on sunny days. Ozone is one of the harmful constituents of California smog and is becoming a feature of long hot spells in England and indeed any area which has pollution, still air and hot sunny weather simultaneously. It is also another greenhouse gas, helping at this low level to add another layer to the blanket round the earth. Research continues on its exact contribution to global warming.

But back to the function of the ozone layer as a filter of ultraviolet light in the upper atmosphere. It is this light which burns human skin, causes skin cancer and cataracts, and damages the tender shoots of growing plants. It also kills some of the plankton on the surface of the ocean, the vital beginning of the sea's food chain, which has been shown to be helpful in removing carbon from the atmosphere. So extra ultraviolet light reaching the planet is bad news for life on earth. But having a thin ozone layer does have one advantage from the global warming point of view: it allows more heat to escape from the earth than a normal ozone thickness would. This minor benefit is far outweighed by the other disadvantages but does offset some global warming which turns out to be just as well. This slight cooling effect of ozone depletion is counterbalanced by another unexpected quality displayed by CFCs – that of being global warming gases in their own right.

Sir John Houghton, co-chairman of the science working group of the Intergovernmental Panel on Climate Change (IPCC), the UN body set up to pull together the science on global warming, says that a CFC molecule added to the atmosphere has a greenhouse effect 5000–10,000 times greater than that of an added molecule of carbon dioxide. This means that despite their tiny concentration compared with carbon dioxide they have a significant impact, and account for about 20 per cent of the total amount of warming by all greenhouse gases. This extra heat is not felt evenly across the globe, however, because of the cooling effect of the thin ozone layer. In the tropics, where there is almost no depletion of ozone, the greenhouse effect of the CFCs is greatest. In mid-latitudes, say over North America or Australia and New Zealand, the effect is neutral. In the polar regions the effect is one of cooling as more heat escapes through the thin ozone than the CFC molecules can trap.

Exactly what is the net effect of having CFCs in the atmosphere on global warming over the long term is not yet clear. For the first time in 1995, because of the efforts towards cutting out ozone depleters, the level of CFCs in the atmosphere was reduced. But partly because of their long lifespan in the air the influence of CFCs will be felt for a long time. Even if all goes well it will be another 50 years before the problems of these synthetic chemicals and the resultant ozone depletion fade into insignificance in the battle to save the planet from ourselves. Meanwhile their effect on climate change will have to be carefully monitored.

One of the problems in banning any useful substance, and as has already been pointed out, CFCs were in many respects a great boon, is that substitutes have to be found and they too may have unexpected disadvantages. The giant chemical companies Dupont in the United States and ICI in Britain spent huge sums producing alternatives. The result was the even more clumsily named hydrochlorofluorocarbons (HCFCs) and hydrofluorocarbons (HFCs). These worked on the same principles as CFCs but were not quite as harmful. HCFCs were still ozone depleters but lasted a far shorter time in the atmosphere. HFCs contained no chlorine and so did not damage ozone – sadly though, they turned out to have a significant global warming effect. So these gases have produced another, although smaller, series of problems. The chemical companies will be allowed to sell enough of these new chemicals to recoup the development costs but they will have to be phased out too.

One interesting aside here was the intervention of Greenpeace in this debate. As elsewhere in this story the pressure groups, or non-government organizations (NGOs) as they are known in the jargon, were constantly in evidence at meetings demanding action. They always want faster phase-outs, more dynamic action than either business or politicians are prepared to accept. The great stumbling block in the replacement of CFCs was that the substitutes took a long time and a great deal of money to develop. Once they were developed, as a reward for doing so, the companies then had to be allowed to make a profit.

Greenpeace rejected that argument on the grounds that complex new substances were not required to run fridges. Quite simple alternative technologies existed which did the job just as well. The organization developed a butane fridge, cheaper to make and just as simple to run, but the resistance of the fridge manufacturers was the next stumbling block. Eventually the organization encouraged a company in the former East Germany to begin manufacturing. With its huge membership in Germany Greenpeace was able to market the fridge to the environmentally conscious and rich European consumer. China, poised on the edge of a consumer society, is now making butane fridges rather than concentrating on the traditional CFC machines.

For the manufacturers and the politicians this campaign to adopt solutions was a new development they had not bargained for.

The environment groups were no longer simply based on pointing out the problems and putting pressure on for solutions, they were suggesting the solutions themselves and if they met resistance were actually prepared to produce the goods. This tactic in the fight to save the planet added a new dimension to both CFC and global warming debates.

One result of the initial priority given to ozone depletion over global warming is that the political responsibility for phasing out CFCs remains with the Montreal Protocol. Because CFCs are dealt with in that forum they are specifically excluded from the Climate Change Convention. The one problem that poses for climate is that one group of substitutes, the HFCs, are not ozone depleters but are global warming gases, but so far the problem is not significant.

# 6

---

# Uncertainty about the size of the danger

ONE OF THE STUMBLING BLOCKS for those anxious to press ahead with action to slow down global warming has been the scientific uncertainties. Scientists seemed to spend more time saying they were not sure of things than pointing to the magnitude of the crisis. It leaves the rest of us puzzled about the size of the predicted rise in temperature caused by the various gases. The more information that has become available the greater the possible problems seem to be. Trying to predict what will happen, and when, is really weather forecasting on a grand scale. With everyone knowing how difficult it is in countries like England to be precise over what will happen in the next week, it is not surprising that scientists are constantly emphasizing the uncertainty of the climate in the middle of the next century.

This problem persists despite the large resources that have been deployed around the world, although some theories can be tested and give greater confidence that previous predictions are correct. Scientists in the United States, Britain and Germany have at their disposal some of the most sophisticated computers yet invented. Countries all over the world have co-operated in measurements of what is happening now and hand over their records as far back as they exist. Temperature, rainfall and wind speeds are constantly monitored all over the globe as part of the international effort to keep track of climate change.

In a perfect scientific world what was actually happening on the ground in terms of real temperatures should be able to be reproduced by computer models. Scientists at the Hadley Centre for Prediction and Research at the Met Office in Bracknell in Berkshire, England, ran their supercomputer through 470 years of weather to see if they could compare theoretical results with real records. The problem has always been that so many reactions are taking place at one time in the atmosphere, even the most sophisticated machine had trouble taking them all into account. In this case what the scientists were trying to factor in was all the pollutants, both those that heat up the atmosphere, like carbon dioxide and methane, and those that cool it down, for example smoke and dust from industrial processes.

One only has to read Dickens' novels or travel to any heavily industrialized region in the Third World to know that even the strongest sun can be almost blotted out by industrial pollution. However, much of the pollution we live with all the time we do not even notice. Throughout the whole of Europe and North America power stations and other industrial processes are constantly pushing smoke into the sky. All these processes, including cars and trucks, produce what are known as aerosols, particles of matter that intercept and scatter a small amount of the incoming radiation, thus reducing the energy that reaches the ground. Some aerosols occur naturally, including dust and evaporated sea salt, but any increase in the number or size of the aerosols increases their blocking power, thereby cooling the climate.

Aerosols are also a key element in the formation of clouds (about which more in the next chapter). The number of droplets in a cloud is determined by the number of aerosols available on to which water vapour can condense. If there are more aerosols available then the clouds tend to be composed of a large number of small drops rather than a smaller number of large drops. The larger the number of small droplets the whiter the cloud appears to be and the more sunlight it reflects back into space. So both these activities of aerosols have a cooling effect.

Whenever fossil fuels are burnt, a number of different gases are released. By far the largest in quantity is our old friend carbon dioxide, but there is also a significant amount of sulphur dioxide. Unlike carbon dioxide, sulphur dioxide is a chemically active gas which reacts with others in the atmosphere. In dry conditions, it changes over a period of a few days to solid or liquid particles of sulphate, increasing the aerosol content of the air. In wet conditions, the sulphur dioxide is drawn into the clouds and absorbed by the cloud droplets to form a weak solution of sulphuric acid.

In Europe the best-known result of this is acid rain, as the sulphuric acid falls back to earth in rainfall. The consequences of this include damage to buildings and paintwork and acidification of soils and freshwater. Large areas of Europe have lost all fish life in some lakes and streams as a result, including much of southern Norway and Sweden. Large parts of southern Scotland, the Pennines and Wales suffer too.

What was not so well known until recently was that these sulphate aerosols had the effect of shielding the earth from the sun and thus had a cooling effect. Is it any wonder that journalists and politicians get confused about the environment? One pollutant can cool the earth but causes acid rain, another (CFCs) punches a hole in the ozone layer, which cools the earth, but at the same time heats it up by trapping infrared radiation.

Another complication is the lifespan of these pollutants. Unlike the gases that cause warming, aerosols are very short lived in the atmosphere, most falling back to earth within a few days. In fact, long-term studies of temperature in industrial areas of Australia showed that the weather was on average marginally warmer at weekends when factories were closed down. This means that to get a true picture of what we are doing to the atmosphere both the heating and cooling gases have to be added in. Until this was done at Bracknell, the assumption had been made that aerosols cooled the earth. This was partly because when only the warming gases like carbon dioxide were taken into account the computer thought the weather was warmer than real measurements found it to be. So when they put both warmers and coolers in, the Hadley scientists were delighted to find that their experiments with the supercomputer mirrored the observed temperature range from 1860 to 1990. This is important because the increase in pollutants from both carbon dioxide and aerosols was quite dramatic over this period.

With this degree of success the scientists felt confident enough to run the computer on in the same sequence to 2050 to see what would happen. To give some idea of the complication of this experiment it was necessary for the scientists to put representations of the atmosphere, oceans, ice and vegetation into the computer. Once that was done it took three months of continuous running of the world's fastest supercomputers to complete the simulation of the climate change. What this showed was that future global warming would be 0.3°C (32.5°F) per decade without sulphates in the atmosphere but 0.2°C (32.4°F) with them.

These remarkable experiments have immense implications. First of all many of the doubts about what is happening are removed. Various regional observations in which temperatures have not gone up as much as predicted can now be explained. Serious consequences are also implied. As we have already seen with CFCs, removing one set of pollution problems affects another. The results of the aerosol experiments explain why much of Europe and North America have not warmed up as much as it was thought they would. In one sense this is a bonus because it gives these governments more time to prepare. But for completely unrelated reasons – the damage that acid rain and other pollution causes – all industrialized countries have programmes to reduce aerosol emissions. The whole of Europe is committed to a timetable into the next century to cut sulphur dioxide emissions to a tiny proportion of their levels in the 1970s. This is very good news for an end to acid rain but if it is successful it means equally that the same areas will be subject to even more rapid warming than they expected and faster than the parts of the world which have no sulphate effect at present.

Indeed scientists have already factored these results into their calculations. They expect the opposite result for parts of Asia, where industrial activity is increasing dramatically. The Chinese, for example, are building one extra coal-fired power station a week. The result is that these areas, which are relatively free of aerosol pollution, will suffer it in far greater measure. For them it may hold back global warming for a time although overall temperatures are still reckoned to rise. The damage caused by acid rain will far outweigh any benefit from a slow-down in global warming.

Synthetic aerosols are not the only cause of dust particles which hold back global warming. Volcanoes inject enormous quantities of dust and gases into the upper atmosphere. There have been records in the past of Icelandic volcanoes creating so much dust in the atmosphere that they caused widespread crop failures in Scotland. In Tambora in Indonesia in 1815 there was a prolonged eruption. It was followed by two exceptionally cold years, and 1816 was described in New England and Canada as 'the year without a summer'.

Those calamities were a long time ago but on 12 June 1991 an event began which enabled scientists to test their theories. A massive eruption that went on for many days at Mount Pinatubo in the Philippines injected an estimated 20 million tonnes of sulphur dioxide into the sky together with enormous quantities of dust. It was one of the most violent eruptions this century.

The importance of the severity of the eruption was that the dust and sulphur dioxide reached heights of 19 km (12 miles) or so above the earth. If the eruption had been smaller the dust and gases would have fallen back to earth in exactly the same way as pollution from factory chimneys. However, these particles did not behave as they would have in the lower atmosphere, known as the troposphere. The dust and fumes from the eruption were caught up in the powerful horizontal winds which circulate the globe in the next layers of the atmosphere, known as the stratosphere, which begins about 11 km (7 miles) above the earth.

The dust spread round the world in a giant ring, causing spectacular sunsets for many months. From the point of view of the scientists studying global warming it provided equally spectacular proof of their theories on dust. The amount of radiation from the sun reaching the lower atmosphere fell by about 2 per cent.

Global average temperatures over the next two years fell by about 0.25°C (32.5°F) as a result. Cold winters in the Middle East were also linked to the effects of volcanic dust. But, however fine the dust, it gradually falls out of the stratosphere into the lower atmosphere. Here, like the pollution caused by human activity, it is rapidly caught in the clouds and washed out in the rain. By 1994 the Pinatubo effect had faded and the surface temperatures, averaged globally, were in the warmest 5 per cent of all years since 1860.

The trend continued in 1995 and scientists were able to confirm at the beginning of 1996 that it was the warmest year ever recorded. Those data were analysed from 1000 weather stations and 1000 ships and buoys, showing exactly how sophisticated the study of climate change has become. The University of East Anglia and the Met Office said it was 0.4°C (32.7°F) warmer than the 1961–90 average. The previous warmest year was 1990, the year before Mount Pinatubo erupted and caused a temporary cooling. That had been 0.36°C (32.6°F) warmer than average. If there are no more volcanoes to interfere with the process then these records are expected to be broken again and again.

One other group of synthetic aerosols is worth a mention – soot. This comes directly from chimneys and some vehicles, for example diesel engines. Being black, soot absorbs solar energy, as opposed to sulphate aerosols which reflect it. This means that soot could have a different effect on climate but so far no one knows quite what it is. When someone finds out, the Hadley Centre will no doubt add it in to their experiments.

So with both global warming and global cooling gases being emitted at the same time by human activity the scientists have a difficult task in telling us what is happening. For what it is worth, they calculate that about 31 per cent of the heat that would have reached the surface is reflected back because of aerosols, clouds and other natural obstructions.

As has already been described, one factor that clearly makes it more difficult to make calculations is the differing lifespans of synthetic gases in the atmosphere. For example the Hadley scientists had to add in to their experiment an atmospheric lifetime of carbon dioxide of between 50 and 200 years, plus an annual increase. Aerosols, although constantly being produced, have an average lifetime of only ten days. They are not well mixed in the atmosphere and have the highest concentrations downwind of industrial regions. So Europe, where there is a prevailing westerly wind, is cooler to the east of industrial areas. The UK's pollution therefore acts as a shield to the continent from the sun as well as giving it acid rain. The particles are normally washed out by rain within a few hundred kilometres of their starting point.

As we have already discussed, these gases also have different properties so their warming or cooling effect is not uniform. The longer they exist the greater effect on the global pattern; hence the need to concentrate on carbon dioxide. In order to factor aerosols into the calculations on global temperature researchers have to assume that industrial production continues to pollute the atmosphere with fossil fuel burning. Given the politics we shall discuss later, this is at best guesswork.

For example, at the moment extensive efforts are being made in Europe to reduce the impact of acid rain from all the inefficient industries of Eastern Europe. If this is a success then it will increase global warming. Similarly, increases in pollution have been factored in for Asia. This seems reasonable for the time being because pollution is increasing daily. With new technologies developing the picture could change radically in the future. For all our sakes we should hope that it does.

Despite all these complications, scientific opinion is that the pleasant average temperature of the earth due to the natural greenhouse effect is changing and will continue to do so because of human activities. In fact, temperatures are beginning to climb more rapidly than at any time in the past 10,000 years.

Already, scientists believe that the temperature has risen between 0.3°C and 0.6°C (32.5–33.1°F) since the nineteenth century. Figures have been checked and rechecked over the last five years and the warming is evident in surface temperatures over land and ocean.

Looking ahead is more difficult since exactly how much more pollutant will be pushed into the atmosphere is not known. The IPCC has made a number of informed guesses. The so-called mid-range emission scenario makes rather optimistic assumptions that civilization makes big efforts to cut pollutants. Even so, global mean surface temperatures increase at a rate between 0.15°C and 0.33°C (32.3 and 32.6°F) a decade when the effects of greenhouse gases alone are considered. If the cooling effect of aerosols is taken into account the rate drops to 0.12–0.26°C (32.2–32.5°F) per decade.

Thus the warming effect of natural greenhouse gases, which is estimated to keep the earth's surface 33°C (91°F) warmer than it would otherwise be, could soon be augmented to 35°C (95°F) or even more. This may not sound much but it is enough to change life on earth faster than it has ever changed during humankind's evolution. It is almost certainly too fast for the survival of many of the myriad life forms on which we depend for our existence. Dr Tim Johns from the Hadley Centre in Berkshire is one of those who have been running the computers through the possibilities to calculate whether the pollutants will cause net cooling or heating. 'As we move into the future we can see that greenhouse gases will increasingly dominate. We infer, based on our assumptions, that global warming will accelerate.' A 0.2°C (32.4°F) increase in temperature a decade is probably twice as fast as more sensitive ecosystems can survive. In mid-latitude areas such as middle Europe and the Midwest region of the United States it has been estimated that a 1°C (34°F) increase in temperature is equivalent to a change in latitude of about 500 km (310 miles). A 0.2°C (32.4°F) per decade rate of change is thus broadly equivalent to a 10 km (6 mile) change of latitude per year. This implies that ecosystems would need to migrate up to 10 km (6 miles) a year in order to maintain their position within their preferred climate type or 'envelope'. Most species are unlikely to cope with this pace of change, particularly long-lived varieties like trees.

From existing observations it is also clear that there has been greater warming over continents between 40 degrees and 70 degrees north. That means from the middle of the United States to beyond the Arctic Circle, and similarly from Spain through Europe to northern Norway and most of China and Russia. Another principal observed finding is that the range in temperature between night and day has reduced. This is because the nights have become warmer. The uneven nature of the warming – it is always going to be greater over land – means that the temperature might rise as much as 1°C (34°F) over 20 years in one location. 'It might mean that types of vegetation and agriculture may become unviable in certain locations where they currently thrive,' Dr Johns said. There will be more of this in the next section on the effects of climate change, but consider for a moment about plants that live on the top of mountains. Where do they migrate to when the weather gets warmer and their niche in the system disappears?

Those with a knowledge of recent British history will know that medieval monks harvested grapes and made wine routinely in Yorkshire. It is perhaps no accident that wine-making has again begun in earnest in England in the last ten years, and Yorkshire again has viable vineyards. Professor Hubert Lamb, of the Climate Research Unit at the University of East Anglia, put together a record of average temperatures for central England for the last 1000 years. There was a steep climb of above 1°C (34°F) in the average temperatures running up to 1200. The average took 100 years to rise 1°C (34°F) and reached just over 10°C (50°F). On Dr Johns' calculations that would have meant grapes would thrive much further north than previously possible. This warm patch lasted for two centuries before average temperatures slumped back to below 9°C (48°F). Grape-growing died out as a practical proposition. The cool spell persisted and grew worse. From about 1400 to 1850 there was what is termed the 'Little Ice Age'. There is evidence that this period of cold extended throughout Europe and North America, and glaciers in other parts of the world suggest it could have extended over the southern hemisphere as well. Exactly what caused these fluctuations is not known since the greenhouse gases remained remarkably stable in the atmosphere during this period.

But back to England and the Little Ice Age. What a difference small temperature changes can make. In England during the seventeenth century it was cold enough for the Thames to be completely frozen over. In the winter Londoners were sufficiently confident of the hardness of the frost to regularly roast an ox on the ice. There are a number of paintings executed at different times of hundreds of people holding parties on the ice with large bonfires to keep them warm. The ice must have been very thick indeed.

Yet the temperature changes which caused these extremes were less than 2°C (36°F). Proper ice ages, where ice sheets render most of Europe and North America uninhabitable, arrive when temperatures drop 4–5°C (39–41°F). Nobody knows with certainty what happens with a rise of 2–4°C (36–39°F). It has not happened in the last 100,000 years and so is beyond our experience, even with all the scientific tricks we have learned. A warmer world may sound pleasant, at least if you live in the mid-latitudes, but the more we find out about the possible consequences the less we like it.

# 7

---

# Cloud cover:
# one of the great complications

ANYONE WHO HAS BEEN IN an aeroplane and passed after take-off from a dull rainy day into bright sunshine above the clouds can have no doubt about the startling difference clouds make to the sunlight reaching the surface. By acting as a barrier between sun and earth they keep the earth cooler. But at the same time the clouds are preventing the heat already trapped beneath them escaping back into the sky. At night clouds act as a blanket, preventing the heat accumulated during the day from escaping back into the upper atmosphere. This is why in the temperate regions of Western Europe frosts are much more frequent on clear nights and cloud is normally a guarantee of several degrees of extra warmth.

As has already been noted, one of the first observations of existing climate change is that although some days are warmer the most noticeable variation is warmer nights. Scientists believe that the most likely explanation is increased cloud cover caused by global warming. Warmer seas mean more water vapour and more clouds. One of the most difficult areas of understanding in grappling with the potential of global warming is the role of clouds. Everyone agrees they are very important but they remain the greatest uncertainty. In fact they are the main weapon in the armoury of those scientists who refuse to accept that global warming is likely to happen. They believe that clouds are the climate's self-correcting mechanism. The fact is that a difference of a few per cent in the total cloud cover of the earth could change net warming into cooling.

What complicates matters further is that there are many different types of cloud, some low and thick, some thin and high, some full of ice or water. Best estimates are that extra low clouds have a slightly cooling effect, because they prevent the sun's rays reaching the earth in the first place, but that high clouds warm things up because they act more like an extra blanket. Current research shows that there are increases in both, but significantly more high cloud which would confirm the warming.

More water vapour in the atmosphere and more low clouds also mean more rain, and so an increase in the cloud cover is good news for dry areas. It will also, of course, in preventing the sun getting through, reduce the evaporation of water already in the ground. Observations show that generally there is now more cloud over Europe, North America and particularly Russia. It also means less sunshine, and it is interesting to note that in January 1996 the United Kingdom had the lowest ever recorded sunshine levels for the month.

Rain appears to have increased over high latitudes in the northern hemisphere, particularly during the autumn. However, there has been a decrease in the rain in the subtropics and the tropics from Africa to Indonesia as temperatures have increased. This is bad news for these regions because the extra warmth means extra evaporation, meaning even less moisture in the soil.

Careful observation of lake levels, particularly in the Great Lakes in North America, gives some idea of trends. Here Lakes Michigan, Huron and Erie have a higher water level than the historic average. There was a major drop in levels due to the intense drought of 1988 but all the Great Lakes except Superior are again above average in height. Levels of the Great Salt Lake in Utah are somewhat similar but the lake has been regulated with the introduction of pumps to relieve the historically high levels of 1986–87.

The Caspian Sea, in the middle of the European/Asian land mass, is the largest closed water body in the world. Its level has been measured systematically since 1837. Unfortunately, human interventions of various sorts have altered the level in recent decades but levels have also risen for natural reasons. Reduced evaporation due to increased cloud cover is one of the explanations. At the opposite extreme are the lake levels in northern hemisphere Africa, where lake levels have dropped drastically. The early 1960s were relatively wet, but between 1965 and the late 1980s Lake Chad shrunk from its highest recorded level to about one tenth of its area. This was due to the drop in rainfall rather than excess extraction for irrigation.

Those who might consider betting on the possibility of a white Christmas should read the next bit. Snow cover, snowfall and snow depth are among the first to be affected by climate change. Long-term lack of data is a handicap here but over the northern hemisphere land surface snow cover extent has decreased between 1988 and 1994 according to figures from a 21-year period of satellite observations. The reduction in snow has been particularly noticeable in the spring. The snow cover in the autumn and summer has also been reduced by about 10 per cent over the 21 years. This decrease has been reflected in an increase in temperature leading to warmer springs. These warmer springs have been borne out by observations of ice melting earlier in lakes and earlier snowmelt-related floods in Canada and California. River ice in Russia, which holds back river traffic, has also been melting earlier. The River Danube, which freezes in the winter, has, on average in the last 20 years, been icing up a week later in the autumn and becoming ice-free a week earlier in the spring. The frost-free season in the north-east United States begins 11 days earlier than 30 years ago.

Since lying snow and ice, like the white clouds, has an important role in reflecting heat back from the surface of the earth, these observations could explain why the increase in surface air temperatures over the northern hemisphere land areas has been more significant in spring than in other seasons. Snow depth tends to be greater because of generally increased precipitation over the region, but in areas on the edge of the snow belt there has been a greater increase in rain rather than snow.

Another straw in the wind of understanding climate change is the measurement of soil moisture contents. Data from the European part of the former Soviet Union taken over 30 years show a general increase in soil moisture from the 1970s to the 1980s consistent with increased rain and reduced evaporation. This shows that factors other than actual temperatures could be more significant in climate change.

All these pieces of the jigsaw are fed into the computer and are useful in that they increase understanding and help us to prepare ourselves to adapt to climate change. Unlike the other gaseous components in the cocktail of climate change, water vapour cannot be controlled, at least not for now. If it was possible, we could make it rain in droughts and turn deserts into gardens. What does seem to be happening is that cloud cover is increasing in some areas, giving the world on average warmer nights. This often means that there is more rain, but almost always where it already rains a lot, for example in northern Europe and in Asia where it is likely to increase the severity of monsoons.

One of the other great and complex factors in calculations of climate change is the role of the oceans. The oceans cover most of the earth yet differ so much in character and temperature, and have such different life forms that it is impossible to model with any certainty what is going on. Currents shift billions of tonnes of warm water every second from the tropics towards the poles. At a different depth cold water is moving back again. These currents are mapped and monitored but exactly what drives them, why they travel at the speed they do, and what might cause them to change course is still not clear. For example, the currents that wash the shores of the British Isles make the climate much milder than it should be at that northern latitude. Coconuts from the tropics have been washed up on Scottish shores and in some parts of the west coast carefully protected but open air gardens can grow plants in a frost-free environment not possible anywhere else north of the Mediterranean.

The range of temperature change both by day and through summer and winter is much less at places near the coast and directly relates to the warmth of this water from the south. If the Gulf Stream, which pushes this water to Europe, were to weaken or turn elsewhere, the climate of Britain would change dramatically. In particular, the winters would become much colder and all predictions about the effect of global warming on Britain, and to a lesser extent the rest of Europe, would change overnight. The ocean current carrying heat towards the west of the British Isles contains as much warmth as is gained from the sun. When this vast current of warmth reaches a certain point it begins to sink. This is because of its extra saltiness, making the water denser or heavier. Like a conveyor belt, the water then turns back on itself deep in the ocean and is carried back south. Scientists studying the climate of the North Atlantic region believe that past climatic changes have caused these currents to change. Melting ice pouring vast volumes of freshwater into the oceans has disturbed the currents in the past

## Great ocean conveyor belt

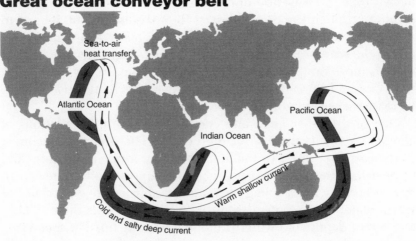

and who is to say this might not happen again, stopping them bringing warmth northwards or diverting them to a different stretch of coast.

Studies of ice cores in Greenland have shown that around 10,700 years ago there was a dramatic increase in temperature of 7°C (45°F) over a period of 50 years. It is thought that this was caused by a sudden resumption of the warm ocean currents after they had previously been stopped by melting ice flooding into the North Atlantic from the land mass of Canada. We have no way of knowing whether melting of extra ice in the Arctic as a result of predicted warming will make these mighty currents falter. If they do, then the climate may suddenly get colder in parts of the North Atlantic while for the rest of the world it gets warmer. Either way there is little we can do about it except monitor the present situation and study the climate records.

Already in this book there has been some criticism of scientists for not being alarmist enough. This possible climate change in the North Atlantic is one which could bring calamity for large communities but we have to accept that the possibility is beyond current records and too uncertain to give reliable predictions. In their 1996 report, IPCC I, the scientists of the IPCC have this to say, not just specifically about warm water currents, but in general about things that lie in store: 'As future climate extends beyond the boundaries of documented impacts of climate variation in the past it becomes likely that actual outcomes will include surprises and unanticipated rapid changes.' That sounds like a pretty scary prediction to me.

The ocean currents and the temperature of the sea are vital in providing some of the energy which creates the weather we experience. In the Atlantic weather patterns are driven by the difference in pressure between subtropical high pressure near the Azores and subpolar low pressure to the south and east of Greenland. The exact positions of these extremes and the difference between the two affects the track of the strong westerlies across the Atlantic. The recent ten-year-long winter dry conditions over southern Europe and the Mediterranean, and the wet spell sweeping from Iceland to Scandinavia, are related to this. So far it is not clear how much is natural variation.

However, an event called El Niño in the Pacific, which is believed to be associated with warming of the oceans, is being studied carefully. Basically El Niño is an invasion of warmer waters into the cooler parts of the Pacific which happens every few years. It appears to cause heavy rainfall, high sea surface temperatures and changes in trade winds in the central and eastern Pacific. The warm surge of water from one part of the Pacific to another can move over normal areas of ocean upwelling and cut off the food supply to fish and sea bird populations. But the event also has much wider effects. It is said to cause droughts in Australia, Brazil and Africa, and unusually high rainfall in central and north-east South America. The effects are felt as far away as North America. These events, once thought to be a rarity, have happened repeatedly in the last 20 years. There is still argument about whether their frequency can be directly attributed to global warming.

One of the predictions from the beginning of the theories on climate change was that there would be more severe wind storms, more droughts and more periods of heavy rain. Scientists are still arguing about this too. As we shall see in the next section on the likely effects of climate change, there are a range of predictions about the damage that extreme weather events will cause. Already many weather records have been broken. For example, research shows that Siberia, a legendary place to avoid getting sent to because of its bad weather, is getting warmer. Trees, in this case Siberian larch, respond to warmer conditions by growing faster. The annual growth record is contained in the tree rings; if they are wider apart the trees had good years. By measuring growth rings, researchers found that since 1901 Siberia has routinely had better summers

## The Coriolis effect

The Coriolis effect is caused by two bodies, such as the earth and a mass of air or water, moving at different speeds. This causes a swirling effect away from the equator. Hurricanes form where there is a strong Coriolis effect.

Fast-moving water along equator

WATER

Currents swirling away from the faster water

## Danger spots

Arrows show main areas of hurricane formation

Equator

than any similar period from AD914 when the subfossil tree rings run out.

It is often claimed that the frequency of tropical storms has increased, and 1995 was a near record year for their severity and frequency. In September the Caribbean was hit by three: tropical storm Iris and hurricanes Luis and Marilyn. October saw Hurricane Opal which hit Mexico, killing 21 people and causing $2 billion damage in Florida. This was followed by Hurricane Roxanne which devastated the resort island of Cozumel before pushing into Mexico. These were just a few of the 17 events in that Atlantic storm season to reach hurricane force, the worst year since 1933 when there were 21. Remember that 1995 turned out to be a record warm year.

Hurricanes are known as typhoons in the South China Sea and tropical cyclones in the Indian Ocean. To develop they require a sea surface temperature of 27°C (81°F) or more. The warmer the sea, the more likely the wind pattern is to develop, and then accelerate. The pattern becomes circular because of a phenomenon known as the Coriolis effect. This is where a mass of water and a mass of air moving at different speeds cause a swirling effect. It looks a bit like water going down the bath plughole. Once started it gets faster and faster and can whip up speeds of 200 km per hour (125 mph) in a

# A storm is born

Hurricanes form over warm, tropical seas when the water temperature is above 27°C (80°F).

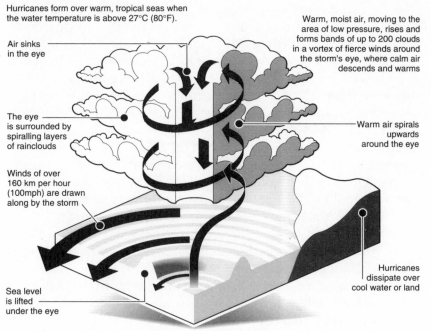

Air sinks in the eye

Warm, moist air, moving to the area of low pressure, rises and forms bands of up to 200 clouds in a vortex of fierce winds around the storm's eye, where calm air descends and warms

The eye is surrounded by spiralling layers of rainclouds

Warm air spirals upwards around the eye

Winds of over 160 km per hour (100mph) are drawn along by the storm

Hurricanes dissipate over cool water or land

Sea level is lifted under the eye

circle up to 650 km (400 miles) wide and is usually accompanied by exceptionally heavy rain. The vortex in the centre of the storm lifts the sea level in the middle. Once the hurricane hits land it begins to blow itself out so coastal regions and islands are most vulnerable. The course of these storms is unpredictable; sometimes they veer out to sea and pick up strength again.

An analogy that has been used about global warming and storms is that as water comes to the boil it becomes more turbulent. The heating up of the world is therefore bound to cause more extreme events. Although most scientists disclaim that there can yet be any link between current experiences and global warming, or at least say it is too early to tell, environmental groups are not so sure. If the tally of extreme events is totted up the evidence is disturbing. For example in 1995 the United States North-East and Midwest had a July heatwave that killed more than 800 people, hundreds in Chicago alone, and more than 1000 tornadoes were reported in the Midwest, the second most active season on record. Moscow and Siberia had record temperatures. Parts of China, India, Bangladesh and Nepal had exceptionally heavy rains leading to flooding, England had both a drought and a heatwave. The increased turbulence of the boiling water analogy does not just apply to big storms. Scientists believe this could also mean sudden switches from warm to cold weather systems and account for the huge snow storms early in 1996 along the Eastern Seaboard of the United States.

But we digress; back to ocean currents. Another reason for their importance is that the sea, with its myriad life forms, is a major absorber of carbon dioxide. Both the water itself and the creatures it contains take carbon dioxide out of the air. If the sea takes enough carbon dioxide out of the atmosphere then it could halt climate change. On the other hand, if the sea takes less then the process of warming could speed up. One of the concerns of scientists is that the colder the water the more carbon dioxide it absorbs.

As warm water is carried towards the poles from the tropics it cools and absorbs more carbon dioxide. Eventually when it reaches the ice it cannot take any more. By this time it is heavier than both the ice and the warm water pushing on the current from behind and so, as we have said, it sinks like a waterfall and, deep in the ocean, reverses direction to head back to the tropics.

A natural phenomenon that clearly demonstrates the remarkable way in which ocean currents move and shows that waters of different temperatures and densities do not mix is the unmarked and invisible line round the southern hemisphere known as the Antarctic convergence. This is where the near freezing water of the Antarctic, so cold that icebergs are still floating about in it, meets the warm water of the South Atlantic and the Pacific. At the point where the two meet the Antarctic waters drop like a huge cascade to the ocean bed and the warm tropical current rides over the top. There is often fog as the cold air of the Antarctic stretches out over the suddenly warm sea. But the most remarkable element is the fantastic fish and bird life on this invisible moving line. The cold waters are full of oxygen and other nutrients so tiny plankton and other organisms thrive. The warmer waters from the tropics are much poorer in food supply. Where the two meet there is a mixing of the water over a few miles. Here, even though it may be thousands of miles from land, there are always hundreds of birds and the whole surface of the sea is full of life as the fish take advantage of the abundance of food that this meeting of oceans brings.

But what effects does all this warming and cooling of the sea have in terms of extracting carbon dioxide from the atmosphere? At present it is calculated that the sea and the organisms that live in the top few metres remove vast quantities of the gas from the air. The carbon dioxide, as in the forests, is incorporated into the bodies of the organisms, some of which are then eaten by other creatures and some of which die and then sink to the seabed.

One of the possibilities raised by scientists thinking of ways to combat global warming is that we could stimulate plankton growth and so get these organisms to absorb more carbon dioxide. In some areas of the ocean plankton growth is inhibited by the lack of iron in the water. One idea was to spread iron filings or a solution of iron into the water and create an algal bloom. This was seen as the stuff of science fiction when it was first suggested in 1988 but experiments have been carried out since. The first attempts were not successful because after a few days the iron solution sank. The second experiment in June 1995 in the Pacific involved using a tonne of iron in a soup of seawater and seemed to have more success. The algal bloom it caused took about 10,000 tonnes of carbon from the air, giving hope to those who believe that human beings can think of a technical fix for global warming. The scientists involved, however, doubt whether even if they could make it work consistently it could make enough difference. The problem is simply too vast.

A flavour of some of the other, slightly less rational technical fixes, all suggested by the US Academy of Sciences, are as follows: placing 50,000 mirrors in space in earth orbit, each of 100 km$^2$ (39 sq miles) to reflect the sunlight back; using guns or balloons to maintain dust in the stratosphere, increasing sunlight deflection; placing billions of balloons in the stratosphere to provide a reflective screen; using high-flying aircraft to trail dust into the atmosphere, or decreasing the efficiency of aircraft engines to produce a thin cloud of soot. All of these may be adopted, of course, but it might be just as easy to cut greenhouse gas emissions.

Anyway, to get away from science fiction and back to ocean currents and to what passes as fact. According to scientists the uptake of carbon dioxide from the atmosphere by the natural process of the gas dissolving in the cooling waters amounts to a staggering 105 billion tonnes a year. However, the reverse process of releasing it back again into the atmosphere occurs when the water resurfaces in the tropics. When the water on the conveyor re-emerges from the depths and is driven to the surface again it warms. As a result it releases a large part of its stored carbon dioxide. At the moment it is calculated that the oceans remove about 3 billion tonnes more carbon dioxide a year than they put back into the atmosphere.

This is a tremendous benefit to mitigating the greenhouse effect. The puzzle is why this should be and whether it will change. Scientists know that warmer water absorbs less carbon dioxide. The oceans are getting warmer, possibly by as much as 1°C (34°F) a decade in some places. This means that the sea could eventually release more carbon dioxide than it absorbs. Then we would be in real trouble.

Calculations connected with what is going on in the sea are complex because it is a constantly moving pattern and the sums involved are so vast. The figures usually have so many noughts on the end that a small margin of error can make a difference of millions of tonnes in the final calculations. Even scientists wonder whether it will ever be possible to describe accurately what is really going on. They know, for example, that the warming of the seas is not uniform. Satellites show that surface warming can be patchy.

On land where plants, trees and soil stay conveniently in the same place there is greater confidence that we can both understand and more accurately calculate what is going on. The point of argument here, which research is still trying to resolve, is how the extra carbon dioxide in the air affects the growth rate of plants. Despite the alarm about the increasing quantities of carbon dioxide in the atmosphere, it is still a relatively scarce gas. This means that plants have to seek it out from the air to carry on their daily business of photosynthesis in order to go on growing. Various experiments have been carried out by scientists which involve keeping plants in a controlled atmosphere, like large greenhouses, where the quantity of carbon dioxide is doubled. This is known as carbon dioxide fertilization.

Studies in both crop and non-crop plants grown in doubled carbon dioxide have shown increased growth responses of 15–71 per cent. Interestingly, the plants respond most positively when the moisture content of the soil is depleted. This is an effect that can be expected in a warmer world.

Some plants respond vastly differently from others. In its 1996 report, IPCC II, the IPCC reports dryly that the total range 'is from a negative response of 43 per cent to a positive response of up to 375 per cent'. For crop plants like soya beans, rice and maize, three of the world's staple foods, it could mean they grow faster and produce more seed. It would mean far less land was needed for the same yields. If the calculations are correct, cotton would double its yield.

So far so good, but it is also clear that this may be a short-lived phenomenon. There are signs that trees are already adapting to the increase in carbon dioxide. Leaves stored in Cambridge since 1750 from common species like oak, beech, maple, poplar and hornbeam are considerably different from those gathered today. Before the industrial revolution began pushing the carbon dioxide level of the air upwards the pores in the leaves through which the trees breathed were far more numerous. The scientists found that in 200 years the average number of the pores, called stomata, had reduced by 40 per cent. Since trees not only breathe but also lose moisture through these pores, such an adaptation helps them to retain water. In all, the trees are better able to cope with greater heat and less moisture but they may not take out as much carbon dioxide as the scientists hoped. In fact other experiments are beginning to show that once plants have got used to increased carbon dioxide concentrations they tend to resume normal growth rates.

Most of the experiments so far have been carried out on crop plants, and obviously the effect on food plants is of vital importance in the future feeding of the world. The effect on trees, however, is likely to be just as important for the long-term future. If the increase in carbon dioxide leads to a faster growth rate in trees and hence a larger quantity of carbon fixed in wood, it will be a substantial bonus. Not only will the world's forests soak up more carbon dioxide, it means that growing large quantities of trees will both be more commercially viable and make a more significant contribution to slowing down global warming. This makes the billion trees promised by President Bush on the 28 million hectares (70 million acres) of America's unproductive lands look a much more worthwhile proposition. Again in England the community forests that are being planted will be soaking up some surplus carbon dioxide.

But the picture on the prospects for carbon dioxide fertilization can be distorted. One of the red herrings which researchers have to contend with is the effect of other pollutants. Leaving aside the effects of acid rain on trees there is evidence that they gain benefits from pollution in their growth rates. This could be the extra carbon dioxide but it could also be the extra nitrogen from fossil fuel burning. In parts of the northern hemisphere with reasonable amounts of heavy industry it is calculated that pollution accounts for an average of an extra 50 kg of nitrogen per hectare (45 lb/acre) per year. This is the equivalent of throwing fertilizer out of an aeroplane. In order to test the theory that this affects tree growth, researchers added this amount of fertilizer to large plots of healthy evergreen and deciduous forests for several years. The result was extra growth, expressed in terms of added carbon storage of between 10 and 20 per cent. However, too much added nitrogen kills trees so the effect of uncontrolled pollution can be devastating. So far no research has been done which put together the dual effect of fertilization from the carbon dioxide with the extra nitrogen. It seems highly unlikely that the trees will grow much faster. Much research needs to be done to understand the processes. Again the progress so far has only been possible because of the great advantage of the research record already in existence in Europe and to some extent North America. The effect of extra carbon dioxide on tropical forests is much harder to gauge. For example, how can we tell if they have already started to adapt? The answer probably is that we shall never know.

# 8

---

# Sea level rise:
# perhaps the biggest shock
# yet to come

WHEN GLOBAL WARMING FIRST caught the public imagination it was the prospect of a rise in sea level that most excited the popular press. In the longer term it is potentially the most damaging consequence of our pollution of the planet, at least as far as human civilization is concerned. After all, if large areas of the most populous and productive areas of the world disappear under the sea it makes the survival of a large proportion of the world's population a problem. The evidence is that the sea level is currently rising quite fast by historic standards. Another rather sobering fact is that there is very little we can do about it; the oceans are very slow to react and it takes hundreds of years to reach equilibrium. That means that even if we managed to correct the imbalance of gases in the atmosphere, sea levels would go on rising for a long time. This is one of the potential surprises and shocks that the UN's scientists refer to.

The problem is that the science surrounding sea level rise is not simple. There are two main reasons for rising oceans: the thermal expansion of water as the world warms and the melting of the ice above sea level in mountain glaciers and at the poles. The 1995 estimate is that over the last 100 years the global sea level has risen by about 1–2.5 mm ($\frac{1}{16}$–$\frac{1}{8}$ in) a year or 10–25 cm (4–10 in) in total. This is, according to tide gauge records, a slightly higher estimate than that made in 1990. It seems odd to have such a wide variation in past measurements but scientists point out that the land on which the tide rises and falls moves up and down too. For

## Sea level rise

Sea level rise predictions in centimetres if no action is taken

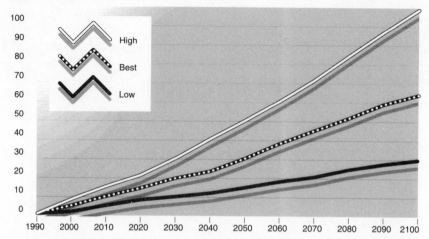

example, the south-east of England is sinking and the north-west of Scotland rising as a result of the last ice age. The weight of the ice pushed Scotland down into the earth's crust and it is still recovering. Recent technological developments in measuring sea level changes from satellites will cut out the problem of distortion because of land movements. We should get both a general and regional picture, because despite the fact that we all know water finds its own level, science tells us that the sea will rise in some areas more than others.

Thermal expansion is said by the IPCC to be one of the major causes of the sea level rise in the last century and it will be in the next. Calculating how much we can expect in the future is an extremely complex problem because water expands a different amount at different temperatures. For example, cold water expands hardly at all when heated. At 5°C (41°F), which accounts for a large area of ocean near the poles, the increase in water volume for each degree centigrade is about 1 in 10,000. In tropical seas, where the temperature is around 25°C (77°F), the rise for each degree is 3 in 10,000. Sir John Houghton calculates that if the temperature of the top 100 m (330 ft) of ocean in the tropics goes up by 1°C (34°F), the increase in depth would be 3 cm (1³⁄₁₆ in). Thus sea level rise as a result of thermal expansion would be uneven across the globe.

The latest estimates say that approximately one fifth to one third of sea level rise over the last century could be because of this thermal expansion, a trend that is bound to continue.

The second cause of sea level rise is the melting of glaciers such as those in the European Alps. Records in Europe are useful because they go back further and with greater accuracy than any others. In other parts of the world scientists can research what might have been happening by calculating the age of the ice and looking at the air and dust fragments trapped in it to see what the climate was like. But despite these techniques there is no substitute for a physical record. In Antarctica, for example, there remain huge doubts about what is happening, partly because there is so little recorded history. It is less than 100 years since humans were able to survive there long enough to measure anything. But we are getting ahead of ourselves. The scientists like to split the ice sheets at the poles, Greenland and Antarctica from the rest of the glaciers and ice caps. This is because the vastness of the polar ice caps and the extreme cold which hangs like a protective veil around them gives them a separate climate and makes them less vulnerable to melting.

In contrast, all over Europe glaciers are said to have been in retreat since the middle of the last century. As they get shorter they also get thinner. It means quite simply that large amounts of ice once sitting on mountain tops are returning to the oceans and making them deeper. An example of this European retreat was when, to the delight of archaeologists, a body appeared in September 1991 on the Otztal Alps. Subsequent work showed that the 'ice-man' had probably died in a snow storm about 3200BC. He appeared to have been on a hunting trip and when he died his body was frozen and incorporated into a slowly moving glacier. If the weather had not warmed in this century and the glacier had not begun to retreat the ice-man would still be in his tomb. As it is we have learned a great deal about his clothing, which included woven grass to keep him warm, and his hunting tools, including a copper axe and copper-tipped arrows. Clearly a wealthy chap. The finding of his body in a slow-moving glacier tends to indicate that the ice level in that valley is less than at any time since he died.

The retreat of all glaciers (outside the polar regions) in the last century is said to have added between 2 and 5 cm (1–2 in) to sea level in that period. That would be evenly spread across the globe. Although the amount of ice in these glaciers is only a small fraction of that at the poles, the rates of accumulation and loss are much more rapid. Many of the smaller glaciers are expected to disappear completely in the next century. Another problem is that the rates of flow of glaciers vary enormously, and occasionally when warmed up sufficiently they can start moving much faster, or surge. Where they run into the sea, as they do round the Arctic, they produce icebergs. Typically they grow longer and thicker in the winter and retreat in the summer so it is not always possible to be precise about the rate of their disappearance or the mass of water they contain.

Although the glaciers of the Alps are the best documented, it has also been noted that glaciers are also disappearing in Alaska and Washington State, and seem to be growing in the Canadian Arctic. For many areas of the world there are no proper records but retreating and expanding glaciers leave many clues for scientists to work on. Suffice it to say that if all these glaciers were to disappear completely as the world warmed, the sea could rise by as much as 50 cm (nearly 20 in).

The real unknown, and the greatest unresolved argument, is over the ice sheets in Greenland and the Antarctic. Again, as in the case of pollutants in the atmosphere, there are competing forces. Logic says that if the earth gets warmer the ice will begin to melt and this is true. Exactly how much melting there will be and how much there has been in past warm periods is a matter that has scientists at loggerheads. More of that in a moment, but it must also be taken into account that in a warmer world, as has already been described, there are more clouds. At the poles this means more snow. In the case of both Greenland and the Antarctic this could considerably add to the thickness of the ice. The Antarctic, which has an ice cap 4 km (2½ miles) thick covering a vast area the size of the whole of Europe, is technically a desert because, despite appearances, it hardly ever snows. This is because its position as a land mass at the bottom of the world gives it a permanent high pressure with only a tiny amount of annual snowfall, and that is mainly at the edges. This means that the ice takes many centuries to accumulate, which gets scientists very excited. By cutting out a vertical core of ice, layer upon layer of annual snowfall can be captured. As each year's snow is compressed, tiny bubbles of air are trapped. Careful study can reveal exactly how much carbon dioxide and other gases were in the atmosphere when the snow fell. The Russians managed to take the deepest cores in the ice in their Antarctic base to find snow that fell 160,000 years ago. It gives an unsurpassed record of the climate over that period.

The ice that is produced as a result of this compressed snowfall is gradually squeezed ever outwards by the weight of the ice on top of the mountains in the middle of the continent. Eventually it falls off cliffs or creeps into the surrounding sea. Huge pieces break off and form icebergs, sometimes 30–50 km (20–30 miles) across. In 1995 a 77 by 35 km (48 by 22 mile) piece of the Larson ice shelf broke off. This ice is so compressed that it is harder than steel and poses a serious threat to even the strongest vessel. Spent icebergs are so heavy that they hardly break the surface of the sea; they are called growlers. Enterprising sailors snap bits off icebergs with hammers for freshwater ice in drinks. The entertainment is to listen when the trapped air is finally liberated after thousands, or possibly millions, of years of being compressed in ice. It makes a delightful pinging noise.

The question is whether the extra snowfall on these great fields of ice as the earth warms will compensate for the increased rate of melting. Evidence from the past is studied for clues. In the ice ages it is clear that when ice covered Europe as far south as southern England and in North America to the south of the Great Lakes, the sea level was as much as 100 m (330 ft) lower than today. The English Channel was dry and it would have been possible to walk from Dover to Calais. In the warm period between the last two ice ages the sea was 5 m (16 ft) higher than now. But it is also clear that large ice sheets create their own microclimate. By reflecting the rays of the sun they avoid heating up. In other words they have an in-built protection mechanism or inertia. Some scientists believe that even if the temperatures in the Antarctic got warmer nearly all the ice would remain in place.

The experts on the IPCC admit that the behaviour of these giant frozen areas is probably the biggest puzzle they have to unravel. They point out that most of the freshwater on the planet is trapped in the two ice sheets of Greenland and the Antarctic. 'Thus, a loss of only a small fraction of this volume could have a significant effect on sea level.'

As has already been mentioned, a giant iceberg was released from the Larson ice shelf in Antarctica in 1995 and it was part of a break-up of both the Larson and Wordie ice shelves. This was in the peninsular area of Antarctica north of the main mass of ice where more than 8000 km² (3000 sq miles) of ice shelf has been lost in the last 50 years. Temperatures in the region have warmed more than 2.5°C (36.5°F) in the last 100 years. In just 50 days in 1995 1300 km² (500 sq miles) of the Larson ice shelf sent a plume of new icebergs into the Weddell Sea. One of the puzzles of these kinds of events is what effect they have on future warming. The Antarctic sea ice, which has no effect on sea level, since it forms and melts each year, is also thought to be reducing. This means, however, that there is less ice and ice sheet to reflect the light. The dark oceans and dark rocks heat up and increase local warming. The ice goes on retreating. If this cycle has begun, was it global warming that kicked it off?

Further south there have also been enormous iceberg discharges from the Filcher and Ross ice shelves, which are part of the main mass. However, as has already been said, human experience in this area goes back less than 100 years so these could be routine events. Even if they are not, then some scientists believe that the increased snowfall in the area would compensate for the ice lost through extra iceberg calving. So despite the dramatic events of the last five years in Antarctica the scientific jury is still out on whether the ice continent is going to cause the sea to rise.

Greenland is regarded with more apprehension. Unlike Antarctica, where the temperature is so cold there is no direct melting of the ice, Greenland loses as much water from melting as from iceberg calving. This means a warming of Greenland could have an immediate effect of thinning the ice, which would not be fully compensated for by the extra snowfall. The IPCC says that observational evidence is insufficient to say whether the current ice sheet is in balance or has increased or decreased in the last century. On balance 'the increase in melting should dominate'. Another indicator is the melting sea ice. Polar bears are said to be in trouble because the sea icefloes on which they live are disappearing. Orbiting satellites give reliable data which has been analysed by Ola Johannessen and his colleagues at the Nansen Environmental and Remote Sensing Centre in Bergen. While the Antarctic sea ice has reduced slightly the Arctic reduction is more significant. The data shows that between 1978 and 1987 the area of sea ice in the Arctic fell by 2.8 per cent. Between 1987 and 1994 it fell again; the equivalent of 4.5 per cent a decade. This will again increase warming and could enhance the melting in Greenland.

But despite these rather doom-laden clues, the role of the great ice sheets still remains one of the great unknowns and many scientists are urgently looking at the problem. If trends are discernible they can be added to the computer models and give us some idea of what the future holds.

One of the scientists involved in the research is Professor David Sugden of the Department of Geography at the University of Edinburgh. He is studying the East Antarctic ice sheet where more than three fifths of the world's freshwater is stored in the form of ice. Some researchers had claimed in an article in 1994 in the scientific journal *Nature* that three million years ago this ice sheet melted. If so, he says, the global sea levels would have been raised about 60 m (200 ft). If this was repeated as a result of the current global warming we would all have to start building arks. The theory about the melting arose because tiny sea creatures called diatoms were found in the interior of Antarctica in glacial deposits. They may have been carried there by glaciers, but their presence in the area at all implied that there must have been sea encroachment into the area. Hence the ice sheet had melted. Professor Sugden, giving a paper at the British Geographers' Conference in Glasgow in 1996, suggested an alternative explanation – that the diatoms had blown there in the powerful Antarctic winds. This would mean that the ice had been stable for 16 million years and it was unlikely to melt unless there was a runaway warming of the planet. He backed this up with the discovery of glacier ice eight million years old under a thin layer of soil near where the diatoms were found.

While the professor was confident that the East Antarctic ice sheet was safe, he was not so sure about the smaller but still very significant West Antarctic sheet. If this melted it could add between 4 and 7 m (13–23 ft) to sea level rise. Unlike the ice sheet on the other side of the continent which is mainly sitting securely on top of a mountain range, the problem here is that the edge of the ice shelf is floating on the sea. The majority of the base rests on the sea bottom, however, held down by its own weight. The sheet is fed by ice streams from the interior, which by glacier standards move quite fast. So far no one knows whether all this will lead to a surge in the ice sheet. It is clear that in the past some smaller ice sheets have disappeared and returned.

When the tide rises and falls, floating ice still attached to the main ice cap is put under enormous strain. When it snaps off, large chunks float away to melt. This is a constant process and as has already been said, records do not go back far enough to be sure of trends. However, large areas of ice around some of the islands in the Antarctic peninsula previously thought to be permanent have collapsed and fallen into the sea in the last few seasons. This has given rise to the as yet unproven fear that this is the start of something more significant.

Once he has settled the argument on East Antarctica, Professor Sugden hopes to add his expertise to the study of the West Antarctic ice sheet. He feels that the greatest danger is that the already rising sea levels will cause a greater area of the West Antarctic ice sheet to float off. If more of the ice floats and the warmer water gets underneath, the tide is likely to work loose even larger chunks of the ice. This is one of the more alarming possibilities.

Even assuming this does not happen, the current estimates for sea level rise are for 15 cm (6 in) by 2030, and about 50 cm (20 in) by the end of the twenty-first century. Because of the in-built time lag inherent in these processes the sea levels would go on rising after that even if the climate had otherwise stabilized. These rises may not seem very much but in the next chapters, when the likely effects are considered, such figures appear to create the potential for a series of large-scale catastrophes.

# PART THREE

---

# THE EFFECTS

# 9

---

# Adapt, sink or
# move to higher ground

Small islands and low-level coastal areas are particularly vulnerable to climate change effects such as rising sea levels and
increases in flooding, coastal erosion, and storm frequency and
intensity, with tens of millions of people at risk. Many low-lying
areas, such as parts of the Maldives, Egypt and Bangladesh would
be inundated and made uninhabitable by a 50 cm (20 inch) sea
level rise.

These are the first two sentences of the scientists' summary of the
effects of sea level rise in the 1996 report, IPCC II.

In fact it understates the enormity of the potential disaster.
There are 36 countries, members of the Alliance of Small Island
States (AOSIS), all of which fear they may be rendered completely
or largely uninhabitable by the middle of the twenty-first century.
Even that is only part of the story. Sir John Houghton, the climate
expert who is chairman of the UK's Royal Commission on Environmental Pollution, in a speech in February 1996 to the Royal
Society in London, reported that large areas of the world could not
be protected, and that particularly vulnerable were the deltas of
large rivers. He added southern China to the list of casualties and
said that the situation in many of these areas would be exacerbated
because, for other reasons, the land was sinking at a similar rate
to the expected sea level rise from global warming. In other words
the effect will be doubled. Two examples he gave were where
movements in the earth's crust are causing the land to sink and

the shrinkage caused by the large-scale extraction of groundwater below many of the world's highly populated regions. 'Substantial loss of land will occur in these areas and many millions of people are likely to be displaced.' For example, 'six million people live below the one metre contour in Bangladesh'.

For some island countries, particularly in the Pacific, any sea level rise is a problem and a rise of more than a few centimetres could spell disaster. These are the islands which owe their existence to coral reefs. Most of them are a maximum of 2 m (6 ft) high. Millions of years ago they were mostly active volcanoes. When the volcanoes ceased to erupt they began to erode and sink back on to the ocean floor. Around the edge of these volcanoes tiny creatures that make up coral attached themselves to the rock and began to grow. As each generation died another grew on top, so as the mountain sank a ring of coral remained. In many cases the original volcano has disappeared completely, leaving a ring of islands with a shallow central lagoon with good fishing. For many this is the perfect holiday destination. The ocean with all its raw power is kept outside by the ring of the islands and inside there is a warm and safe lagoon for scuba diving and swimming off silver sands.

Global warming and subsequent sea level rise will bring many problems to these holiday paradises. First of all the extra heat of the water can cause extensive bleaching of the coral, killing the tiny creatures that constantly replenish the reefs. Secondly, the coral, hampered by the bleaching effect and extra storm damage, may not be able to grow fast enough to keep up with sea level rise. With islands only 1–2 m (3–6 ft) high the sea can wash over them in a storm. This is normally a once-in-a-lifetime experience, and although the devastation can be great and in some cases people lose their lives it is rarely enough for people to abandon their homeland. The direct wave erosion of the coastline, partial inundation during storms, and seawater intrusion into fresh groundwater, which is often the island's main supply, are likely to be common factors in making life more difficult in the future. Some scientists have concluded that the most susceptible nations will be those 'composed entirely of atolls and raised coral islands, which will be devastated if projected rises occur'; consequently 'such states will cease to be habitable islands'.

More detailed studies of individual island states show that responses could be different and it is possible that the action of the

# Sinking small islands

Some of these countries are so low-lying they could be inundated in the next century by a rise in sea level and storms. Others, among them the world's most popular tourist spots, will lose valuable coastal land and main centres of population will be drowned. These countries are demanding urgent action to slow down global warming.

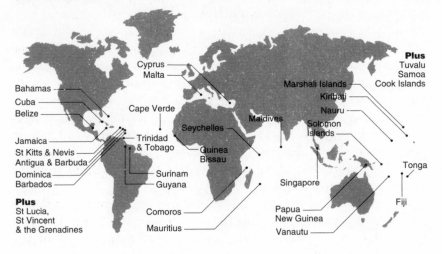

sea will throw up enough coral debris to replenish and rebuild some islands; in other places corals could grow fast enough to keep up with the rising sea level. Some nations, however, are likely to end up as 'uninhabitable sandbars in the middle of the ocean'. Even with the most optimistic picture, the combination of sea level rise and the possibility of more frequent storms makes life on these islands look increasingly chancy, particularly those in the tropical regions where there are more intense storms and sea level rise is expected to be greater because of the thermal expansion of the water.

Even if the island states themselves survive the damage which climate change inflicts, it is more than likely to wreck their economies. Many small islands in the Maldives are less than 1 m (3 ft) above sea level, and will be reduced to sandbars, restricting usable dry land to the larger, more populated islands. But the wealth of the Maldives, and perhaps this is why it gets more headlines than other similar but less well-known destinations, is based on tourism. In 1991 alone the industry netted $94 million, 74 per cent of the country's foreign exchange earnings. It is a figure that has since risen. Among the Caribbean islands tourism is also vital for economic survival. Even though the islands are not threatened with inundation their beaches are. For at least a dozen islands, including the Bahamas, loss of tourism spells financial disaster.

Although the concept of whole countries disappearing under the waves is a headline grabber, the disturbing fact is that a large percentage of the world's population lives on low-lying land next to the sea. Estimates vary, but it is generally thought that about one third of the world's crop-growing area and the homes of one billion people are at risk from a 1 m (3 ft) sea level rise. For example 85 per cent of the world's rice production takes place in South, South-East and East Asia. About 10 per cent of this production is located in areas that are considered to be vulnerable to sea level rise, thereby endangering the food supply of more than 200 million people. Such a sea level rise could create 50 million environmental refugees – more than three times the number of refugees from all causes in the early 1990s.

Current estimates of global sea level rise represent a rate that is two to five times higher than that experienced over the last 100 years. But locally and regionally the rate and magnitude of sea level changes will vary substantially due to differences in ocean conditions and the vertical movements of the land. Extra rainfall, storm frequency and intensity also need to be taken into account, because they could increase the chances of flooding, erosion and saline intrusion. A key point is that the concentration of human activities on the coast has made the situation far worse than it would otherwise be. As the authors of the 1996 report IPCC II put it: 'In the past estuaries and coastal wetland could often cope with sea level rise, although usually by migration landward.' But they add that the fact that people have built cities, roads and sea defences in the way prevents that happening. In general sea level rise would damage tourism, freshwater supply and quality, fisheries and fish farming, agriculture, human settlements, the insurance, banking, oil and gas industries, and human health.

At present in an average year an estimated 40 million people in the world have the misfortune to be flooded due to a storm surge, most of them in the developing world. Ignoring possible adaptations and population growth these numbers are expected to double or treble in the next 100 years. The scientists conclude that protecting all these areas from the sea is impossible and planned retreat may be the only option for large areas.

Some of our longest inhabited, most populous and fertile regions are already at or in some cases below sea level. In Europe large parts of East Anglia and the Netherlands are below the average high tide, sometimes for miles inland. Keeping these places free of floods is already an expensive business. Even with the best engineering, huge financial resources and the best weather forecasting these areas are still sometimes subject to flooding. If sea levels rise by as much as a metre (3 ft) and storm surges in the North Sea grow more frequent, these areas become more vulnerable. In the Netherlands design standards for new seawalls were raised 66 cm (26 in) as a result of projected sea level rise. In eastern England the increase for the same reason is only 25 cm (10 in), but it is a much more sensitive political issue the other side of the North Sea.

In very wealthy countries sea defences are seen as a solution to the problem but the capital sums involved are vast. In Japan, for example, an analysis of the structural measures needed to protect 1100 ports, harbours and neighbouring areas from a 1 m (3 ft) sea level rise is put at a staggering $92 billion. That is $63 billion for raising the height of port facilities above the sea and another $29 billion for breakwaters, jetties and embankments to keep the water out.

There are other areas of the world which have similar problems but do not have the kind of resources required to keep the sea out. The flood plain of the Yellow River which grows huge quantities of food for China is in some places 6.5 m (20 ft) below the level of the river. Its banks have been built up by countless generations of peasants trying to stave off the floods.

Not far away in China's Pearl River delta, where 15 million people live, the vast majority have homes up to 2 m (6 ft) below the high-tide mark. Sea embankments are either of silt or clay and there has been a continual extension seawards by embanking fishponds. The belt of protecting mangroves between the delta and the sea has virtually disappeared, leaving no protection from waves crashing on the banks. The honeycomb of fishponds affords some protection but no thought has been given to locating villages on raised ground to protect the populations. The emergency plan involves successive waves of soldiers from the Red Army plugging the gaps with sandbags.

Sea level rise may not seem much of a threat to vast countries like the United States but there is a heavy concentration of population and economic activity in low-lying coastal zones. Almost 50 per cent of the population lives within 32 km (20 miles) of the coast. Some cities, like Miami, are built on what was once swamp. Miami is really only just above the water level. The freshwater aquifer floats on the salt beneath and the foundations of the buildings stand on a crust which holds them just out of the water. A rise of a few centimetres and Miami will return to the swamp from which it came.

In other places the dangers of sea level rise are compounded by other problems brought about by human activity. For example the Nile delta is sinking. This fertile plain, where much of Egypt's food is grown, used to increase in size every year with the annual flood. The silt which was deposited with each flood made the land slightly higher and pushed the delta further out into the Mediterranean. Then huge dams were built upriver to regulate the flow and produce electricity. The mud which was once flushed down every year now silts up the dams and the delta, starved of its benefit, sinks and its seaward edge is gradually washed away in the storms. If the sea levels rise 1 m (3 ft) then 10–15 per cent of Egypt's productive land will be lost and up to 10 million people will lose their homes. In a country which has a fast-growing population and a shortage of housing this will make already difficult social problems acute.

The Mississippi delta in the United States is also sinking. It was recently in the news for massive flooding, which may be partly caused by climate change, but also by the levees used to control the water. These squeeze the water into narrow channels so it cannot reach its natural flood plain, and so when it does burst its artificial banks the devastation is far worse than it would otherwise be. Like all deltas it is sinking under its own weight into the mud below, in this case at the rate of 1 m (3 ft) a century. Without human interference this would not matter because the replenishment of the delta would come from the natural floods. However, the silt, along with the floodwater which carries it, is no longer allowed to spread out to maintain the height of the delta as it once was.

Bangladesh is the country most often cited as at risk from the combination of artificial and natural disasters. It also has a massive and growing population and very little cash. Even without interference, large parts of Bangladesh would flood regularly as the monsoon rains fill to overflowing the rivers that drain down from the Himalayas. Add to that sea level rise and storm surges, and normal flooding could get far worse. With 80 per cent of the country built on the deltas of great rivers like the Ganges, Brahmaputra and

Meghna, floods are a normal part of life. With half the country less than 5 m (16 ft) above sea level, catastrophic flooding gets more likely with an extra few centimetres of height added to the Indian Ocean. There is a third important factor at work here: deforestation. Upstream of Bangladesh, once heavily wooded mountain slopes have been stripped bare. This means that the trees that once held back the melting snow and the rains no longer do so. It means that the water can rush down all at once and cause even worse floods than before. But it also means more silt. More silt means higher land and larger deltas. Each year Bangladesh gets bigger and more land is colonized.

There are controversial plans to try to control the rivers in Bangladesh by creating channels and barriers to prevent flooding. This would reduce the damage in some areas but the water would have to go somewhere and it may make the problem far worse in other places. This is another Mississippi scheme in the making, only much bigger and potentially much worse. Where the silt goes and whether it can be controlled and made to deposit itself in places to protect the delta from the sea is a matter of guesswork. It is doubtful that it will ever be enough to counteract the sea level rise, particularly since storm surges are expected to worsen. In any case, if the water is diverted from the interior of the delta to limit the damage done by flooding it will also cause a sinking of the land. Already groundwater is extracted to irrigate during dry periods. That in itself causes the land to shrink further and the delta to drop. It also has the long-term effect of sucking in saline water from the sea. Best estimates at the moment are that 17 per cent of Bangladesh will disappear under the sea. Whatever happens, life in Bangladesh is not going to be easy.

Even in countries with plenty of resources by comparison, and a relatively small problem, it is still hard to manage. In the south-east of England where, as we shall see in a moment, sea level rise is greater than in most places, estuaries are hard to manage. Saltmarshes, so valuable to bird life, and as barriers to storm surges and wave damage, would naturally migrate inland as the sea rises. However, centuries of flood defences in the form of banks stand in their way in a process called 'coastal squeeze'. The choice has to be made between losing the beaches and marshes, and building the banks higher to keep out the sea, or breaching the banks and allowing the sea and the buffering marshes to re-establish themselves further inland.

Although it goes against the grain of 2000 years of keeping out the sea, a series of limited experiments have begun called managed or 'planned retreat'. This is also the policy in Australia, particularly on sparsely occupied coasts. Coastal residents are allowed to live on the shore but are not permitted to defend it from the sea, being obliged to retreat before the sea's natural advance.

The role of these coastal zones for wildlife has hardly rated a mention so far as a result of the worries about human settlements and crop-growing facilities disappearing, but coastal zones are the most biologically diverse and productive of all ecosystems. For example, of the 13,200 known species of marine fish, almost 80 per cent are coastal. The shallow coastal waters and habitats like mangrove swamps and lagoons are vital for fish breeding. It is on these breeding facilities that many stocks of fish depend. These are vital for providing protein for the poorer human communities that inhabit the tropical coastlines of the developing world. These coastal habitats are already disappearing at an alarming rate because of human developments but sea level rise will add to the problem. It could severely affect the fish populations, which are already under pressure because of overexploitation. This is a point that the scientists make over and over again about sea level rise; it compounds the environmental problems that humans have already created for themselves.

What preoccupies politicians, however, and in terms of economic catastrophes catches the attention of most voters, is the fate of cities. This is not just because most people now live in cities but because nearly all wealth and the means of producing it is tied up in investments, and what they need to grow is stability. But sea level rise and the violent storms that come from the sea threaten some of the world's biggest cities. The problems stem from the fact that people build houses and communities in locations that look good to live in. The better the choice of site the bigger the city eventually becomes. It might originally have been because it was a natural harbour like New York, Sydney or London. In time these cities develop way beyond their original purpose and become homes to people who have no interest in the sea or the port.

In London, for example, the river is now of virtually no commercial importance, even though it once made its inhabitants rich through trade with the rest of the world. But even though the river is redundant as far as the commercial life of the city is concerned, it is still there, and in the context of sea level rise it constitutes a threat. Although it looks docile enough, the Thames has the potential to destroy much of the commercial heart of the city which was built around it. This was acknowledged as long ago as 1953 when there was a terrible flood in the east of England, which also devastated a much larger area of the continent, particularly Holland. It took a long time for the reaction to be turned into construction but eventually the Thames Barrier was built. This is basically a gigantic series of steel doors which are kept raised above the water to allow shipping through but can be lowered when a combination of high wind and tide threatens to bottle floodwater up in the Thames.

The construction decision had nothing to do with global warming, although it did have something to do with sea level rise. It has been known for many years that the south-east of England is gradually sinking. It may seem a tiny amount each year, around 2 mm ($\frac{1}{16}$ in), but it soon adds up, especially when the sea is rising at a similar rate. At the same time as the south is sinking the Scottish mountains are rising, so keeping them level with the sea rise. This tilting of the British Isles is a correction caused by the weight of glaciers being removed from Scotland after the last ice age. Incredible though it may seem, since 10,000 years have passed since the glaciers retreated, the mountains are still rising, and tilting the south-east into the sea.

The flood disaster of 1953 was caused by a combination of high tides and high winds which pushed a surge of water down the North Sea. A gale from the east coast of Scotland roared down towards the English Channel and caused the tide to build up to record levels. Flood defences all along the British and European coastline were overtopped in a single night and thousands of people lost their lives. It was the worst flood disaster ever known in Europe.

Everywhere along the coasts of Europe sea defences were repaired and raised in order to avoid a repetition. Parts of the lower Thames, particularly Canvey Island in Essex, were devastated. Central London escaped the flood although it was a very close thing; the high-water mark was only 2 cm (1 in) below the top of the barriers on the Embankment. With so much of London, particularly its commercial heart, so close to the Thames, the potential for a multibillion pound disaster shocked the authorities. As an insurance against such a thing happening again the Thames Barrier was built. The barrier, which basically blocks the incoming tide and prevents river levels getting to danger level, is lowered as a precaution whenever the combination of high tides and high winds threatens a storm surge. With global warming adding sea level rise to that already being experienced in the south of England, the Thames Barrier looks an even better investment, although recent estimates say that even that may not be enough to stop the sea if the worst case of high tide and storm came together. If the predictions are right, the storm surges which caused the 1953 flood are likely to be more frequent. This is not regarded as a distant threat. There is already a 24-hour, seven-day-a-week year-long network of control centres all over England ready to warn the public of flooding. These operate with an early warning system from the Met Office and were run initially by the National Rivers Authority which has been taken over by the Environment Agency. When an alert is given, the police have orders to evacuate people from their homes to higher ground.

In the sophisticated countries of Europe and other developed parts of the world the threat of such disasters can be monitored and to some extent prepared for. In less affluent and well-organized countries the first inkling people get of flooding is often when the water starts pouring in the front door. But even for the well-off countries having barriers does not solve all the problems of sea level rise. The aim behind building the very expensive Thames Barrier was to save the most valuable part of London from a potential flood. But what about those areas outside London which cannot hide behind that barrier or afford such a foolproof device against the tide? Of course all these have built their own barriers of one type or another. In fact at present all of the United Kingdom and other European countries which have towns and villages in low-lying areas have sea defences. They vary from simple earth banks to solid concrete promenades or, more recently, a whole series of sophisticated barriers. Often banks of boulders and sand dunes have been created to break up the waves and absorb the shock before the sea pounds on the seawall itself.

Building a strong seawall is more than a simple wish of human beings to save their homes and businesses from the sea. The whole value of property is based on the theory that the land it is built on will remain stable on the same spot. The property is insured against being struck by lightning, civil disorder, wind and flood damage on the basis that they are most unlikely. The business and commercial world is built on the idea of stability. Whole cities are built on the basis that the coastline will not move, and the sea will stay where it always has. A vast industry, insurance, is based on the assumption that what has happened in the past will happen again in the future, but not very often. To put it another way, if flooding occurs on average once in 100 years then people need to insure against it. Because it is a rare event the risk is small and so the annual premiums are also small. When the flood comes and affects some of its customers the insurance company has enough money in its reserves to pay out. This system works because the insurance company can calculate the risks from past information of previous floods. As we shall see in the next section, things start to go wrong when the past records cannot be relied on, and still worse if it becomes obvious that a city is sinking, the sea is rising or both. Insurance companies will not go on accepting low premiums when the risk of a crippling payout is rising all the time. It raises the spectre of a city which becomes uninsurable – and who will invest in businesses that cannot be insured?

It is not just insurance that makes the difference; huge sums of money are invested in tourism, and beach resorts all over the world would find their greatest asset – their beaches – disappearing. This scenario varies from the extreme example of the Maldives, where we have already said the entire country disappears, to resorts where the beach will have to be repeatedly or even continuously replaced artificially. It is worth noting that about 20 per cent of the world's coasts are sandy and backed by beach ridges, dunes or sandy deposits. Studies have shown that over the last 100 years 70 per cent of the world's sandy shores have been in retreat and only about 10 per cent advancing. The rest are regarded as holding their own. Tourism is important in many countries with suitable coasts, but apart from losing beaches, they may suffer from extra storminess, rain and cloud cover. Indirect effects like losing freshwater quality and supply, already a limit to expansion in southern Europe and India, and human health impacts are another inhibiting factor against development. But more of that in a later chapter.

It not just beaches which are threatened by sea level rise. For example the Broads of eastern England, a series of lakes at sea level just inside the coastal fringe, have a tourist trade worth £3 million a year. Barriers to protect them are just being raised in a massive and expensive scheme. The storm-buffering function of the Terrebonne wetlands in Louisiana in the United States is valued in terms of storm damage avoided. If they continue to disappear at their present rates of 3–4 m (10–13 ft) a year the increased property damage is expected to be as much as $3 million a year. The total monetary value of all the tourist and other activities in the Ecuadorian Galapagos National Park, the area instrumental in sparking Darwin's revolutionary theory of the origin of species, is estimated at around £100 million a year. All this could be lost because of sea level rise.

The subsidence caused by groundwater withdrawal can be substantial. Even without sea level rise 2.1 million people in Japan have sunk from living above sea level to below it in a few years simply by extracting the groundwater below them for drinking. Bangkok in Thailand is another example. Between 1940 and 1960 the height of the city above sea level dropped about 3 mm (⅛ in) a year. Since then, when increased groundwater pumping began, it has dropped at 20 mm (5 in) a year, making it extremely vulnerable.

In areas where the climate will get wetter, and extra winter periods of rain are already being experienced in Europe, the combination between heavy rain and sea level rise will cause extra flooding. The water, held up by higher tides, backs up rivers which burst their banks, again a process not helped by our interference with the river system which generally increases the flow of water towards the sea. One place where sea level rise from extra rainfall is already evident is the Caspian Sea, bordered by Russia, Kazakhstan, Turkmenistan, Iran and Azerbaijan. In a sense it does not belong in this chapter because it is a salt lake rather than a sea, with a salinity about one third of that of the sea, but its levels are rising and causing damage to coastal settlements in exactly the same way as general sea level rise would do.

Its main water supply comes from the River Volga and rainfall over the last 30 years in its catchment area has risen 10 per cent compared with the previous 30 years. Under global warming predictions this is bound to increase causing further rises in water level, swamping the ports and fishing villages on its shores. The water level has already risen 1 m (3 ft) in 15 years. If this is not to continue rising, a surplus from the River Volga of up to 200 km³ (50 cubic miles) of freshwater a year will have to be diverted. Across the Ural Mountains is the Aral Sea, shrinking fast because of massive irrigation programmes which have diverted the rivers that kept it alive. If that idea is not possible another possibility is for a 2000 km (1200 mile) tunnel through the Caucasus Mountains to Turkey and the water-starved Middle East. In today's political climate that sounds like an impossible dream but we have the technology if we have the willpower. The Caspian Sea is remarkable in that it is 28 m (92 ft) below general sea level so there is no chance of releasing its flow into the Black Sea. Unless some solutions are found, the rising of the Caspian Sea will become an unfortunate dress rehearsal for what is to happen to the rest of the world.

# 10

---

# Water wars and
# the battle for survival

THE WORLD FACES A CRISIS over lack of freshwater even without climate change. Already 40 per cent of the world's population have water shortages, and with rapidly rising populations demand increases all the time. In a statement to the press, Ismail Seageldin, vice-president of the World Bank, said in August 1995: 'Many of the wars of this century were about oil, but wars of the next century will be over water.' The Bank were reporting that 80 countries around the world were experiencing water shortages which threatened their agricultural industry and health. That took no account of shortages yet to come as a result of climate change, so graft on to that already grim picture the prospect of lower rainfall, high temperatures, more evaporation and more extended droughts in most of the countries already suffering water shortages, and a crisis can rapidly turn into a series of disasters.

If countries are dry but rich, like some of the Middle East oil states, then freshwater can be created with desalination plants run on cheap oil. Nine countries in the Middle East are already forced to import water or desalinate it to survive. But for most of the world that is simply not an option; rain, wells, rivers and melting glaciers are the providers of water needed for survival. Tensions will build between communities and nations over the water crisis because many countries rely on the same rivers for their basic needs – drinking and irrigation. One country's extra irrigation scheme might remove a lifeline for a community downstream for whom the supply will dry up. A war would therefore not be about economic advantage, as in oil wars, but about survival itself. First, let us look at a few figures and examples of the countries affected to give some idea of the scale of the problem.

The IPCC scientists' 1996 summary on climate change, IPCC II, says in part:

> Water availability currently falls below 1000 cubic metres per person a year, a common benchmark for water scarcity, in a number of countries (e.g. Kuwait, Jordon, Israel, Rwanda, Somalia, Algeria and Kenya) or is expected to fall below this benchmark in the next two or three decades (e.g. Libya, South Africa, Iran and Ethiopia). In addition a number of countries in conflict-prone areas are highly dependent on water originating outside their borders (e.g. Cambodia, Syria, Sudan, Egypt, Iraq).

The World Bank figures were stark. Djibouti, a tiny state at the mouth of the Red Sea, has the lowest reserves of renewable water supplies in the world, just 23 $m^3$ (5000 gallons) per person per year. Kuwait has 75 and Malta 85. Altogether there were 18 countries in the early 1990s which had less than the 1000 $m^3$ (220,000 gallons) of water per person per year. Comparison with other more rain-rich countries helps. For example the UK has 2090 $m^3$ (460,000 gallons) of renewable water resource per person per year, and still contrives to have shortages some years, as does the United States with 9913 $m^3$ (2,181,000 gallons) per person, but there the water is not always in the right place for the population's needs.

By 2025 the World Bank estimates that there will be 34 countries which have dropped below the 1000 $m^3$ (220,000 gallon) benchmark. By this time Djibouti will have slumped to just 8 $m^3$ (1760 gallons) of renewable water supplies per person, and Kuwait will be down to 42. A whole swathe of countries across the Middle East and Africa will be desperate for water. These are the areas which scientists think are most likely to both heat up and lose rainfall as a result of climate change. Both areas have been subject to droughts in the last ten years, but then this is a drought-prone region.

Perhaps the classic example of a group of countries depending on one river is provided by the Nile. The Nile flows through some of the driest regions of North Africa and is vital for the irrigation

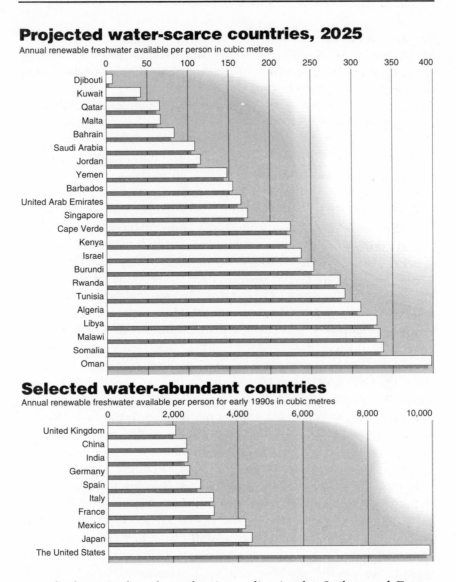

## Projected water-scarce countries, 2025

Annual renewable freshwater available per person in cubic metres

## Selected water-abundant countries

Annual renewable freshwater available per person for early 1990s in cubic metres

on which agricultural production relies in the Sudan and Egypt. In fact, 97 per cent of all Egypt's water comes from the Nile and 95 per cent of that comes from eight countries upstream through which the river and its tributaries run. In 1959 an agreement was made between the Sudan and Egypt over sharing the waters of the Nile but it was not signed by the other seven countries, many of whom might be tempted to improve their agriculture by irrigation at Egypt's expense. Egypt, which has a powerful army – at least compared with its upstream neighbours – first warned in 1985 that it would not countenance interference with its water supply. Ethiopia, frequently struck by drought and famine, has periodically considered using water from the Blue Nile for irrigation projects but has been ordered by Egypt not to do so. The Sudan could also benefit from more irrigation but only at the risk of attack from its far more powerful neighbour. The other countries upstream – Kenya, Rwanda, Burundi, Uganda, Tanzania and Zaire – could all use more water, and three of them appear on the World Bank's list of water-impoverished countries.

There are two more examples in the same region, one of which has already been part of a war and a second which is already leading to tension. One outcome of the 1967 Arab–Israeli war was the occupation of significant parts of Jordon's territory, including the headwaters of the river which gave the country its name. Jordan lost a considerable amount of its own freshwater supplies as a result and Israel gained considerable quantities. Syria and the Lebanon also have a claim on the water. Settling disputes over which country has access to these water supplies in the future is a vital part of the Middle East peace process.

Perhaps the greatest potential for trouble in that already tense region is from the rivers that flow from Turkey – the Tigris and the Euphrates. Turkey controls the flow with 33 dams but downstream both Syria and Iraq depend on the Euphrates for survival. In the early days of the Persian Gulf War there were discussions at the United Nations about the possibility of cutting the flow of the Euphrates River and so denying Iraq's water supply. This would be a little like forcing an otherwise impregnable medieval fortress to surrender by starving the garrison. The Syrians later complained that the Turks had used 'the water weapon' in 1990 by threatening to restrict river flow into their country if they did not stop giving support to Kurdish rebels operating in southern Turkey. Turkey denied it had done so but clearly it was a very sensitive issue.

In southern Africa the question of sharing the water from the Zambezi is a cause for uneasy discussions between Zambia, Zimbabwe, and South Africa, all three countries with water shortages and large increases in population. This is one of the areas which scientists have identified as already appearing to suffer from the effects of global warming. The five warmest years this century in southern Africa have all occurred since 1980 and the warmest

# Temperature rise

South African temperatures, 1901-1994, temperature anomaly (°C)

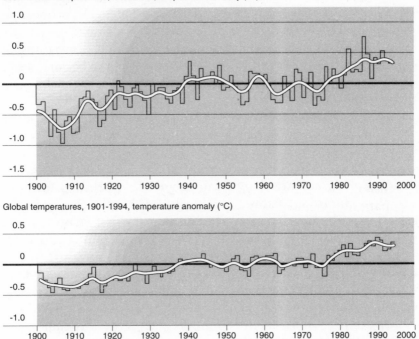

Global temperatures, 1901-1994, temperature anomaly (°C)

decade has been 1985 to 1994. Rainfall in the region has always been variable and droughts have always occurred from time to time. The last 20 years, however, has seen a trend towards reduced rainfall and during the 1990s two serious droughts have occurred. The current population of the region is around 128 million and is growing at an average rate of 3.1 per cent a year, representing a doubling of population every 24 years. There has been a threefold increase since 1950 when there were only 42 million inhabitants. The current population remains 69 per cent rural but there is a constant migration to the towns and cities caused partly by the inability of subsistence farmers to survive the successive droughts. In 1992, 20 million people in the region (about a quarter of the population excluding South Africa) were in need of food relief because of drought.

This is clearly one of the areas of the world where the stress of climate change is going to add to the already difficult problems. It is ironic too that this is one of the poorest areas of the world, which burns least fossil fuel, and has had virtually no impact at all in causing global warming. In fact the issue seemed so remote from Africa's problems that 4 of the 11 states in the region did not join the Climate Change Convention in 1995, a far higher proportion than anywhere else in the world. The World Wide Fund for Nature (WWF) believes that these nations need to start lobbying for their own survival. The organization is active in the area because some of its high-profile campaigns are aimed at saving the wildlife in this vast and beautiful region. The larger animals like rhinoceros and elephants are under heavy pressure from poaching but more fundamentally from human population growth, droughts and a series of wars.

A WWF sponsored report introduced by Adam Markham, the director of the climate change campaign, at a workshop for politicians in Harare in October 1995, showed that the climate in the region was already changing. He said that the first victims would be poor people living in rural areas. An example was the 13 years of low rainfall and crop failures in western Zimbabwe which had led to boreholes drying up, making it impossible to maintain livestock herds and forcing women and children to walk for a whole day to collect water for their families' needs. The Kariba Dam, which produces most of the country's electricity, was running at 14 per cent capacity. Animals were suffering too, with large species, such as hippos, unable to survive without regular supplies of food and water. South Africa's Cape of Good Hope was particularly vulnerable. Warmer conditions, already beginning and predicted to get worse, would reduce the numbers of wild animals that rely on grazing. He said the region was naturally dry but increasing aridity would not only exacerbate problems of feeding the region. It would also hinder industrial and urban development by placing great strains on the economy.

One of the conclusions of the climate scientists who compiled the WWF report concerned the suitability of the area for growing maize, a staple crop. At present almost all of southern Africa north of Namibia and Botswana is potentially useful for maize cultivation, but under the predicted climate change over the next 30 years the maize-growing areas will shrink to a few small pockets in central Africa. This same climatic shift to warmer, drier conditions will badly affect all the larger animals. Elephants and cattle alike will struggle to survive. The best chance for local farmers is to switch to smaller, tougher browsers like goats. In all, the chances of the region supporting itself are reducing, leading to the possibility of the mass migration of the poor to cities or across borders in a bid to survive. The findings must have come as a shock to the politicians. For them, climate change is no longer a problem to be dealt with when everything else is sorted out, because the predicted changes, which are nearly all adverse, will be happening in the space of one lifetime. WWF hopes Africa will soon be demanding action at the conference table.

While in southern Africa there has been a welcome improvement in the political conditions in the 1990s allowing for greater co-operation on climate change, the potential for arguments over scarce water resources remains, and there are examples in every region of the world. In India, the flow of the Ganges, vital for 300 million farmers, is being disturbed by deforestation in the foothills of the Himalayas. India blames Nepal, and Bangladesh complains about the effects of India's barrage just inside its borders and says that its neighbour is taking more than its fair share of water in the dry season. Further west the Indus is vital for Pakistan but two of its biggest tributaries rise in India and are used for irrigation in the Punjab, another cause for dispute between two unhappy neighbours.

Elsewhere in Asia the giant Mekong is shared between Laos, Vietnam and Thailand, all rapidly industrializing and irrigating. So far there is no agreement about plans to dam the river and how to regulate its flow. In the United States there are disputes between states about how to use rivers which flow between them, and more seriously with Mexico over the Rio Grande. Mexico wants more water for its domestic needs but the United States says no. Even in Europe, where countries are bound together in one political bloc, there can still be tensions. The rivers that flow from Spain into Portugal do not have enough water for both countries to exploit them to the maximum, and Portugal has protested at plans for new dams on the Spanish side of the border.

While the potential for wars over water may grab the headlines, the problems of lack of freshwater, or in some cases too much of it, are going to affect the entire planet. It is clear that many of the health effects of climate change will be caused by epidemics of water-borne diseases. This is dealt with in a following chapter, but the existing failure of people all over the world to treat supplies of freshwater as a valuable resource are making a bad situation worse. One of

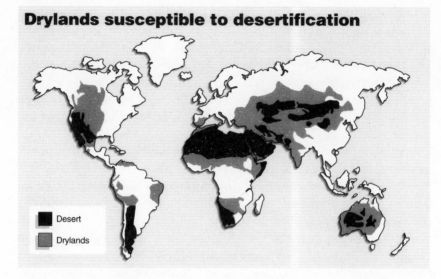

**Drylands susceptible to desertification**

Desert

Drylands

the key points in adapting to climate change is to make maximum use of these diminishing supplies.

Water resources are the key to where the world's population has settled. As the 1996 IPCC report, IPCC II, put it:

> Agriculture, hydroelectric power production, municipal and industrial water demands, water pollution control and inland navigation are all dependent on the natural endowment of surface and groundwater resources. Civilizations have flourished and fallen as a consequence of regional climatic changes and many 'hydraulic civilizations' were formed around the need to control river flow. From 1940 to 1987 global water with- drawals increased 210 per cent while the world's population increased by 117 per cent. A global water resources assessment for the year 2025 suggests a likely population forecast of 8.5 billion, an increase of 55 per cent on 1990 and with globally balanced economic growth, world water use may increase by 70 per cent.

Obviously perhaps, the most dramatic effects will be felt in the regions that are already uncomfortably dry. According to the IPCC report approximately 30 per cent of the earth's land surface is desert or semi-desert, with almost 5 per cent receiving less than 70 mm (3 in) of rain a year, 11 per cent less than 100 mm (4 in) and 18 per cent less than 120 mm (5 in). That is less than a quarter of the rainfall in

# Freshwater

Global freshwater vulnerability of the population in millions

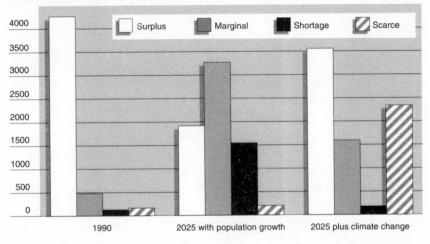

the driest part of England, East Anglia, which gets more than 500 mm (20 in) a year.

Populations in dry lands – classified as having less than 75 growing days a year because of water shortages – are said to total 800 million. These include a large part of northern Africa, the Middle East and extend as far east as northern India, Australia and chunks of southern Africa and North America. One of the predictions of climate change is that these regions will grow larger. The Sahara desert and those dry belts in the Middle East are expected to move north and some have suggested they will leap the Mediterranean, making Greece, Sicily, and southern Spain even hotter and drier in the summer than they are now.

Plants that cling on through heat and drought by special adaptation will suffer greater stress. Climate change will make life tougher because even the small amounts of water available will evaporate more quickly, reducing the growing time available. One of the problems of drier soils is erosion. This is made worse when people remove the natural vegetation to make room for crops. Then, partly through the action of wind, as happened in the dust bowl in the United States in the 1930s, and partly by the rapid washing away of topsoil when it does rain, the land can rapidly degrade, losing the fertile top layers of soil. Higher evaporation rates can also lead to more rapid salt and acid accumulations which badly affect vegetation. This is particularly noticeable in areas of high irrigation, and once highly productive parts of India have been severely damaged. Rising sea levels add to the ingress of saline water into groundwater which, if added to the topsoil or seeping from below, can destroy farming land.

Even in temperate regions a small increase in temperature can have marked effects on the need for more freshwater to irrigate crops. Research in countries that will be worst affected and most need irrigation has still not been carried out; industrialized countries with their own interests at heart have naturally first worried about their own needs. Studies in Britain show that an increase in temperature of 1.1°C (34°F) by 2050 may result in an increase in spray irrigation demand by 28 per cent over and above the extra 75 per cent already anticipated as being needed to meet growing demands. Similar calculations in Poland resulted in an estimated 12 per cent increase in irrigation demand for a 1°C (1.8°F) increase in temperature. In the United States a 2°C (36°F) rise in temperature would require a 20 per cent increase in irrigation. Even where studies made the assumption that there would be more rainfall, some extra water was needed for irrigation because of the increased temperatures.

Industry, which uses vast quantities of water, may find its activities curtailed by shortages. For example, several French nuclear power stations were forced to operate well below design capacity during the drought of 1991 because there was not enough water in the rivers to cool the plants. Hydroelectricity, which relies on regular water flows, is also likely to be affected, sometimes with some benefit. In New Zealand, for example, warming of the weather will give more regular flow and there will be less demand for heating. In California, the reverse is true because the peak water flow will be in January, the period of least demand, and lowest in the summer when the demand for air-conditioning is greatest.

Little work has been done on river navigation but low flows can obviously affect traffic. During the drought of 1988 in the United States the Mississippi flow reduced to a point where river traffic was unable to carry the grain harvest. Increased temperatures will lengthen the duration of the navigation season on rivers which freeze annually. European rivers such as the Danube, Dnieper, Don, Lower Volga and those entering the Black Sea freeze up 2–3 weeks later these days than they did 100 years ago. Improved conditions for shipping companies on the Great Lakes in North America are also expected. They currently have to stockpile cargoes in the winter, but in 50 years' time the Great Lakes may not freeze at all. In temperate regions rivers which normally freeze for a period each winter are expected to become ice-free, and even in colder regions the ice season is expected to shorten by a month.

The effect on wildlife in Africa has already been mentioned but the inhabitants of streams, ponds and lakes are also very vulnerable, not just to volumes of water, but to temperature change. Many species can only survive and thrive in cold or cool waters. Again some studies have been done, most of them in the industrialized countries, where fish are commercially bred for food and recreational fishing is a big industry. The findings are daunting. Apparently, many fish will simply die out as the temperature of their habitats rises. Naturally they would like to move northwards to more suitable climes but in many cases this will clearly not be possible. Human intervention in restocking suitable fish further north seems the only solution, but this would require the introduction of all the other creatures that make up the food chain, a difficult operation. Whole sections of the game fishing industry will disappear.

# 11

---

# Forests, mountains and disappearing snow

FORESTS AND THEIR FUTURE aroused heated debate at the Earth Summit
and they certainly seem to have a special place in the human heart
and the environmental argument. Despite the massive clearance
of trees over many centuries for farming, building and firewood,
forests still covered a quarter of the land surface of the earth
in 1990. The future for forests looks grim, however. Although
extensive efforts are being made to manage forests more carefully
by selectively cutting timber and allowing regrowth, they are still
disappearing at an alarming rate and climate change is likely to
strike a number of additional severe blows. Yet forests are vital
in the future battle to save the climate because they store about
80 per cent of the above-ground and 40 per cent of the below-
ground carbon, which would otherwise be floating around in the
atmosphere as carbon dioxide, and making the earth heat up even
faster than it is already.

Although forests, particularly in temperate regions, are often
managed or even farmed, the majority are still relatively undisturbed
by human influences. Outside their use for wood they are valuable
for tourism, wildlife and the protection of water resources. Forests
harbour the majority of the world's biodiversity, about two thirds
of all the species on earth, only 1.4 million of which have been
discovered and described, the rest being still buried deep in the
forests. This is especially true of the tropics, where about half the
species are lurking under the canopy and, according to the IPCC II

report, 'they represent indispensable, self-maintaining repositories of genetic resources'.

Forests are split into three broad groups: tropical, temperate and boreal – the fir trees of the north. All three are sensitive to climate change and forest distribution is generally limited by water availability or temperature. A sustained increase of 1°C (34°F), well below what is expected, will cause substantial differences to the distribution of tree species. The shorter lived, rapid reproduction species will have a greater chance of adaptation, but even then it is thought that in many regions forests may disappear completely.

Tropical forests are more vulnerable to clearance by humans than to climate change at the moment but whichever wipes them out it will lead to irreversible loss of biodiversity, in other words to a rash of species extinctions. The opportunity for extra forest growth caused by increasing carbon dioxide in the atmosphere is expected to be greatest in the tropics, but so far the long-term nature of this effect is still not substantiated. There is some evidence that trees may simply get used to it and go back to a normal growth pattern. Tropical forests also rely heavily on large quantities of water so decreases in rainfall and soil moisture, especially where it is already marginal, will have a greater effect than a rise in temperature.

Temperate forests have similar problems to tropical ones, particularly if moisture varies, but it is here that changes in tree type are likely to be most marked. Research has shown that if climate change is rapid then conditions may become unsuitable for trees to complete one or more stages of their life cycle. For example, pollen and seed development require minimum heat and are sensitive to frosts. Seedlings are particularly vulnerable to short-term droughts, saplings to the presence or absence of sunlight, and mature trees to the availability of soil water during the growing season. The scientists conclude that tree populations could appear to be quite healthy, while losing their ability to complete their life cycles. Seedlings from trees well-suited to a particular site will grow into adults in climates that may be unsuitable in 50 to 100 years; conversely, adults able to grow in an area in 50–100 years' time must grow from seedlings that may be unable to survive in current climate conditions at those sites. Temperate forests have one slight advantage in that they occur mostly in developed countries, that study them closely, and are prepared to intervene to keep them healthy, even if that means helping them to migrate north to better climes.

Boreal forests are likely to have the hardest time in terms of climate change because they are in the regions where the temperature is expected to rise faster than anywhere else and settle at a much higher average than now, as much as 4–5°C (39–41°F) above current temperatures. Increased fires and outbreaks of pests and diseases, particularly on the southern boundaries of these forests, is expected to reduce their ability to store carbon, as well as their general health. Temperate forests are expected to take over their niche. However, the northern limit of these forests is dictated mostly by temperature. Forests are usually absent when the mean temperature of the warmest month falls below 10°C (50°F), in other words it is not warm enough for them to grow. This means that under climate change boreal forests will have the chance to colonize land that is currently tundra, which will itself have a chance to move north, as we shall see later in the chapter. Even so, the scientific consensus is that boreal forests do not do well in climate change and reduce in area and quality of trees, and so in future will generally release more carbon dioxide to the atmosphere than they store.

Mountainous regions, which also contain a lot of trees, have attracted little research by comparison, but climate change will have a significant effect on them too. As a result it will affect the lifestyle of the human communities that live in both the mountains and the foothills and plains below, especially those that rely on the rivers for farming and industry. Mountains account for about 20 per cent of the surface area of the world and are found on all continents. They usually have sensitive plant and animal colonies and have more frequent extreme weather and natural catastrophes. Climate change is likely to enhance both these effects. Mountains also act as natural barriers to climate, attracting higher rainfall than the surrounding land and often creating what is called a rain shadow behind them. Because their height makes a significant difference to temperatures and the soils are often poor and rocky they frequently have a range of flora and fauna that is vastly different from that of the plains below. They also have different bands of vegetation up their slopes in relatively short distances. Abrupt changes in temperature, which are forecast for mountain ranges, will disrupt the present distribution of vegetation, ice, snow and permanently frozen ground, all of which contribute to keeping mountains stable.

The limited possibilities available to species needing to migrate to favourable locations make mountains vulnerable islands in a sea of surrounding vegetation. Because climate change models are not precise enough to make predictions for small areas of the earth's surface detailed results for mountains are not available. However, some studies have been done for the European Alps which show that in the expected doubling of carbon dioxide levels, winter precipitation will increase by as much as 15 per cent in the western Alps, but ominously for the skiing industry, temperatures will rise by as much as 4°C (39°F). This means that much of the time it will rain rather than snow, and even if it does snow it will melt more quickly. One possible effect is that the snow season will start later. Analysis of satellite data from the 1980s and early 1990s has shown that lowlands around the Alps experience about 3–4 weeks less snow cover than they have historically had. Sensitivity studies show that areas below 1500 m (5000 ft) react quickly to small changes in temperature, especially in the southern part of the French Alps. One of the suggestions to mitigate the effect is the use of large artificial snow-making machines, but these are expensive to run and could be environmentally damaging.

In the United States, ski industry losses from projected warming is estimated to be as much as $1.7 billion annually. In some countries ski resorts could relocate at higher altitudes but in Australia and Scotland the sport and industry is likely to be eliminated altogether. The change in snow cover will also affect the scenic value of mountain regions and may reduce general tourism both in winter and summer.

Apart from the effect on skiing, reduced snow cover will cause the alpine water to run off more quickly in the winter and spring, leading to drier soil in summer and greater fire risk. In the Alps, according to computer models, summertime rain will decrease and July temperatures will go up as much as 6°C (43°F) higher than at present. This is well above average global warming predictions. Other mountain regions would suffer similar problems. Places which are already well-known fire hazards, like the coastal mountains of California, the Blue Mountains of New South Wales in Australia and Mount Kenya, and all the ranges on the fringes of the Mediterranean Sea, would have increased risk of fires. This would put Los Angeles and the Bay Area in California, Sydney in Australia and the coastal resorts of Spain, Italy and southern France at an even greater risk than at present.

The impact of losing glaciers has already been discussed in relation to sea level rise. Their shrinkage seems to be the most obvious sign of gradual warming over the last 100 years, a process which is set to accelerate. By 2050 a quarter of the glacier mass could have disappeared and up to one half in a century. An upward shift of 200–300 m (650–980 ft) in the line of glaciers is expected and an annual reduction in thickness of 1–2 m (3–6 ft). This means some mountain ranges will lose their glaciers altogether. Although the records in the northern hemisphere are better, this is clearly a worldwide phenomenon. In New Zealand some glaciers have retreated 3 km (2 miles) this century and the surface of the Tasman Glacier has thinned by more than 100 m (330 ft).

Along with a removal or retreat of the ice and snow, and in many cases increased rainfall, comes a greater danger of rockfall and landslides. Glaciers leave deeply cut troughs in mountains which are unstable once the ice retreats. Large rockfalls are often caused by groundwater seeping through joints in the rock. Scientists believe that the number of both large and small rockfalls would increase as a result of climate change. Damaged communications and destruction of property would increase. At risk are communities in the Alps and Rocky Mountains but also large urban areas close to mountains like the South American Andean cities, Hong Kong and again Los Angeles, all of which have spread into high-risk areas.

The increase of rockfalls and mudslides has considerable effect on rivers and on the population downstream because of the mud and debris carried with the water. Sediment-loading, as it is called, can have serious effects on hydroelectric schemes by wrecking the machinery. On the other hand less snow and glacier ice will influence the seasonal river flow by reducing meltwater production in the warm season. In some cases, as glaciers melt it will increase summer flow but the river will all but disappear when the glacier shrinks to nothing.

One river mentioned in an earlier chapter, the Rio Grande in the United States, has been studied for run-off patterns. A predicted temperature rise of 4°C (39°F) for the Rio Grande basin increases winter run-off from the snow fields from 14 per cent of the total flow to 30 per cent. As a result, summer run-off is reduced from 86 to 70 per cent, affecting irrigation water and electricity supply – another reason for tension between the United States and Mexico.

Although global warming can have many disadvantages, as these last two chapters amply illustrate, it can also have benefits at least for some people. Better navigation in normally ice-bound rivers and lakes has already been mentioned. A lot of Arctic sea routes presently kept open by ice breakers will be navigable most of the year. One very large difference, which many will welcome, is the northward movement of snowlines by as much as 5–10 degrees in latitude. That is more than the distance north between London and Moscow. The snow season would also be shortened by more than a month. At its maximum in late winter snow covers almost 50 per cent of the land surface of the northern hemisphere – but that is less and less likely to happen in the future. In North America snow accumulation and melt models show a 20 per cent reduction in snow cover for every 1°C (34°F) rise in temperature. On this basis, climatic warming would cause a 40 per cent decrease in snow cover over the Canadian Priaries and a 70 per cent decrease over the Great Plains.

However, more frequent periods of open water for rivers, lakes and seas will produce greater snowfall downwind even if it lies for shorter periods. This is expected to be important near Hudson Bay, the Great Lakes, the Barents Sea and the Sea of Okhotsk. In alpine areas, as already discussed, the snowline could rise by 100–400 m (330–1300 ft) and rain, falling instead of snow late in the season, may cause avalanches, but this is less certain.

For those who live in the far north, particularly those who exploit the world's natural resources for the rest of humanity, global warming could bring calamities as well as benefits. Permafrost, that layer of ground towards the Arctic Circle and on mountain ranges that is permanently frozen, covers 25 per cent of the global land surface. In many areas on the edge of the Arctic it shows ominous signs of melting, and in doing so threatens to destroy the foundations of many oil and gas pipelines, railways, roads and buildings. Permafrost has what is known as an 'active layer', that is the top few centimetres or even metres which melts in the summer. Much of the permafrost layer is left over from the last ice age and because of its extremely cold temperature it takes centuries to melt right through. The top several metres can become 'active' each summer, however, creating lakes and river flows where there were none previously. Engineers have assumed in the past that areas of permafrost were like rock and would never melt and so treated the ice-bound ground as solid foundation. As a result pipelines, airstrips, community water supplies and sewage systems are in danger. The permafrost zone in China carries more than 3000 km (1900 miles) of railway and 13,000 km (8000 miles) of roads, all built on the assumption that there will be no melting. Present permafrost engineering assumes that melting will not be worse than in the warmest year in the last 20. Since every year in the Arctic fringe is expected to get warmer, this standard will have to be rapidly revised.

In these last two chapters some of the bigger effects of the availability of freshwater or lack of it, and the reaction of natural systems to the extra heat, have been discussed. These range from perhaps the more obvious reactions of the hottest, driest parts, to some of the slightly more surprising effects on the colder regions. What the human race does to the purity of that water and its effect on our health comes next.

Top: *Hurricane Hugo, a severe tropical storm with winds of 120 km per hour (75 mph), approaches Charleston and the Florida coast in September 1989. These storms, also called cyclones and typhoons, already cause serious damage and loss of life. They may become more intense and frequent because of climate change. (©Crown copyright.* Reproduced with the permission of her Majesty's Stationery Office.)

Bottom: *One of the Maldive Island chain in the Indian Ocean. These 'paradise' islands are expected to be the first to suffer because of sea level rise with dozens of islands being reduced to uninhabitable sand bars as the sea washes over them. (Rod Dymott)*

Top left: *The left bank of the River Rhine in Cologne in December 1993. Europe is expected to have more extreme weather in the next century, particularly more heavy rainstorms which cause flooding like this.* (Popperfoto/Reuter)

Bottom left: *Sudan, normally a parched country, gets too much water in August 1988 as the Nile bursts its banks. Poor people suffer most from disasters like this which are expected to become more frequent.* (Jeremy Hartley/Panos)

Above: *The Thames Barrier, designed to protect London from severe flooding. It is raised to hold back the combination of high tide and storm surges in the North Sea. Sea level rise shortens the odds on a serious flood overwhelming even this impressive defence.* (Graham Turner/The Guardian)

Top right: *Britain's biggest coal-fired power station, Drax, which produces 10 per cent of the UK's power and large quantities of carbon dioxide that add to global warming.* (Kippa Matthews/Pick of York)

Bottom right: *Boats thrown onto a highway by Hurricane Marilyn in the Virgin Islands, September 1995. The aftermath shows the power of the storms. Marilyn was the fourth hurricane to hit the Caribbean in as many weeks.* (AP Photo/Mark Wilson)

Below: *Bangladesh, threatened by sea level rise and already suffering regular monsoon flooding. People use a pump connected to a well below the flood in an attempt to find clean water to drink to stave off disease.* (Shahidul Alam Badal/Drik/Oxfam)

Above: *Yorkshire, normally one of England's wetter counties, suffered a severe drought through 1995 and 1996. Here Ripponden Reservoir was at only 12 per cent of capacity in October 1995.* (Popperfoto/Reuter/Bob Collier)

Right: *The destruction of forests releases millions of tonnes of carbon dioxide into the atmosphere. Ten per cent of the Amazon rain forest in Brazil (shown here) was cleared in the 1980s.* (Mark Edwards/Still Pictures)

*Icebergs break away from the giant ice sheets of Greenland and Antarctica. If these ice sheets shrink then the sea level will rise dramatically, swamping cities and productive farmland.* (British Antarctic Survey/P. Cooper)

# 12

## Malaria, plague and other unpleasantness

THE HEALTH EFFECTS of climate change are divided into two categories by scientists: the direct effects, that is, being killed by heatstroke or drowned, and the indirect effects, that is, getting a nasty dose of cholera, malaria or plague which would otherwise have been avoided. Extra heat means that tropical diseases can move north and south to trouble previously immune populations in temperate regions. Mostly, however, as with almost every other bad effect of climate change, it is the already poor and overcrowded who will suffer most.

Although the indirect effects will ultimately be more serious, the so-called direct effects are surprisingly marked. Apparently human populations have a heat threshold above which the number of deaths escalates. In the summer of 1995 in Chicago and surrounding districts 600 people are said to have died directly as a result of a short summer heatwave. Scientists predict that this effect will escalate and that in large cities in North America, North Africa and East Asia there will be several thousand extra deaths annually. Since the number of very hot days is expected to double because of the extremes brought on by climate change, the number of people who will die directly as a result increases everywhere – that is as soon as the local tolerance threshold is passed. Work done in Atlanta, Georgia in the United States shows that an average of 78 extra people die each year in heatwaves compared with the normal death rate. By 2020 this number is expected to rise to 191 and by

2050, 293. Research indicates that populations in warmer climes have a higher heat tolerance threshold and so there is a possibility that people in temperate zones will be able to acclimatize to global warming. Even so this will only slightly reduce the increases in death rate.

The most vulnerable groups are the old and very young, who find it difficult to adjust to heat, and those with heart and lung illnesses. There is an association here too with air pollution. Hot weather is often associated with poor air quality and scientists believe that the two factors are likely to have interactive impacts on health. There is a suggestion that most of the deaths are among groups that are already vulnerable through age or ill-health and so the heatwaves and air pollution only hasten their deaths – this may not be much comfort if you are old or ill. One beneficial effect should not go unremarked. In countries like the United Kingdom the winter death rate is much higher than the summer rate. This is mainly through old people succumbing to the cold because of poor insulation and poor housing stock. One study has estimated that in England and Wales 9000 fewer winter deaths would occur annually by the year 2050 with a 2.5°C (36.5°F) increase in winter temperatures. But overall, say UN scientists, extra heat is expected to kill people in greater numbers than those saved by the warmer winters.

The second, more obvious, direct effect of climate change is extreme weather events of other sorts. We are all familiar, even if only through television pictures, with the damage that tropical storms can inflict. These, combined with sea level rise and tidal surges, will continue to kill many people in coastal communities and on small islands. Their severity and frequency is expected to increase, although researchers are still divided about this. It is also important to remember the effect of drought. In early 1991 4.3 million people were facing starvation in north-east Africa as a result of drought. Without massive food aid many more would have died. These droughts are expected to become more common and widespread as a result of global warming.

Death through starvation could fairly be described as a direct health effect. As stated earlier, however, it is the indirect effects that are going to cause the most difficulties. And, getting back to the subject of the last two chapters for a moment, it is clear that freshwater supplies for drinking, the presence of water-borne diseases and lack of adequate sanitation are already linked. Climate change only makes these already bad problems worse. As the World Bank pointed out in a statement to the press in 1995, one billion people lack access to clean drinking water in the developing world and 1.7 billion do not have adequate sanitation facilities. Lack of clean water and proper sanitation kill people in large numbers already, but also can trigger economic problems which make it difficult for communities to recover. In the first ten weeks of the 1995 cholera epidemic caused by contaminated water in Peru, losses from reduced agricultural exports and tourism were estimated at $1 billion. This was more than three times the amount Peru had spent on providing decent public water supplies in the previous ten years.

About 95 per cent of the world's sewage is poured straight into rivers and other water flows, where it is joined by growing amounts of industrial waste. It is these same rivers that many of the poor have to rely on for their only source of water, unless they buy it from water vendors. In many Third World cities where drinking from the river would be fatal, vendors charge more than £1.50 or $2 for a cubic metre (220 gallons) of water, about ten times more than is paid by those who live in houses connected to the city water supply. The health benefits provided by better water and sanitation services were demonstrated in industrial countries in the nineteenth and twentieth centuries. When services were improved the impact on health was revolutionary. For example, life expectancy in French cities increased from about 32 years in 1850 to 45 years in 1900, with the timing of these changes relating closely to the improvements in water supply and sewage disposal. In the early twentieth century some cities in the Ohio river valley in the United States used untreated water, while others treated theirs. Over a ten-year period death rates from typhoid fever were constant in the untreated water cities, but declined by more than 80 per cent over the same period in the settlements using treated water.

In storms and floods the quality of the water treatment systems and their isolation from sewage is vital for the health of the community following a disaster. Extra warmth, which allows the bacteria to breed, and the existing health of the population make a difference to the size and effect of the subsequent disease outbreak. Salmonella poisoning, already a worldwide problem, is significantly worse in warmer weather. In 1983 in Bolivia, in the flooding following the severe weather caused by the El Niño event (where extra warming of water in the Pacific causes abnormal currents and disturbs wind and weather patterns) there was a 70 per cent increase in salmonella infections, mainly in children.

The IPCC scientists say that climate change will have significant effects on the distribution and quality of surface water, including increases in both flooding and water shortages. Lack of water concentrates organisms, impedes personal hygiene, and impairs local sewerage, all of which increase the risk of diarrhoeal infections including cholera and dysentry epidemics, particularly in developing countries. Many of the organisms that cause these diseases can survive in water for months, especially at warmer temperatures. An increase in diarrhoeal disease in impoverished communities with poor sanitation is expected. If flooding occurs this will be worse, with refugee camps particularly at risk.

New research has shown that the cholera organism can survive in the environment by sheltering in the mucous outer coats of various algae and plankton. These organisms increase in warm weather, particularly when fed by extra nutrients from sewage and fertilizers. Increases in coastal blooms of algae, which have been noted during hot spells over most of the heavily populated regions of the world, could become extra cholera carriers. The algal blooms also contaminate fish and shellfish with toxins which can cause paralysis. Higher temperatures also increase the problem of food poisoning by enhancing the survival and proliferation of bacteria.

Studies done in Britain during the hot summer of 1995 showed that salmonella poisoning was significantly increased as a result of cross-infection in abattoirs, particularly in poultry. If this can happen in what are supposed to be optimum conditions in an industrialized country how much worse could it be in less hygienic surroundings? Cockroaches, which are all established tropical importations, have produced a population explosion in Britain during recent mild winters. Like flies and rats, cockroaches enjoy warmer weather and spread diseases. Housedust mites, now the main suspect for the rapid increase in childhood asthma in many temperate countries, will thrive on increases in both temperature and winter humidity expected with climate change.

Many of the world's major infectious diseases are caused by parasites or viruses transmitted to humans via such creatures as mosquitoes and ticks, known to the medical profession as vectors. Cold-blooded vectors are extremely sensitive to temperature and humidity. Diseases thought likely to increase in both range and number of cases are malaria, dengue fever, African sleeping sickness, river blindness, elephantiasis, Chagas' disease, and a form of skin and internal organ disease without a catchy name, known as leishmaniasis.

Of these, malaria is the headline grabber because everyone has heard of it; we are all familiar with the mosquito, and it is the disease most likely to benefit from climate change. The point is that at the moment the types of mosquito that carry malaria only thrive in hot countries, but with an increased global temperature of a couple of degrees, an irritating insect could potentially become a carrier of a killer disease almost worldwide. The World Health Organization estimates that 300–500 million people are currently infected with malaria each year, of which two million die annually. The majority of these are young children and 90 per cent of them live in Africa. This pattern of incidence is set to change dramatically, with malaria spreading widely with the predicted increase of 3–5°C (37–41°F) over most of the land surface. There is more than one type of malaria mosquito and most need winter temperatures of 16–18°C (61–64°F) to survive. However, some can hibernate in more sheltered sites. One of the first signs of general global warming will be the spread north and south of malaria belts and also to higher land.

In Africa this is already significant because the mosquito cannot survive above the height of around 750 m (2500 ft), which leaves the large urban areas of Nairobi in Kenya, and Harare in Zimbabwe, malaria-free. This is likely to end. Overall there is a projected increase from around 45 per cent to 60 per cent of the world population living within the potential malaria transmission zone by the latter half of the twenty-first century. Populations previously unexposed to the disease will lack natural immunity and so fatality rates would be high, at least to start with. Mosquitoes are also responsible for the transmission of another killer called dengue fever, a severe influenza-like disease, which, if left untreated, can kill 15 per cent of those infected. As with other mosquitoes, the species responsible for carrying dengue gets more active the warmer the climate, and an increase of 3–4°C (37–39°F) could double the transmission rate of the virus. The habitat of this type of mosquito is restricted to areas with a winter temperature above 10°C (50°F) but there is evidence that it is already spreading both north and south from Central America.

This kind of scientific information tends to put a damper on some of our natural reactions to climate change and the feeling that if in places like Britain hot sunshine could be guaranteed and grapes harvested for wine, life would be just fine. By 2030 Edinburgh is expected to have the current climate enjoyed by the Midlands, and southern England is predicted to have the climate of the Loire Valley in France. This sounds attractive until you realize that this would make it warm enough for malaria to thrive in southern England. Mosquitoes capable of carrying a milder form of the disease already exist there and it is getting close to warm enough for more dangerous species to flourish. With increasing air traffic it is believed that the opportunities for introduction are high and that it would only be a matter of time before Sussex and Kent became malaria areas. This is not as unlikely as it sounds. Malaria used to be widespread in Britain in the Middle Ages and a more dangerous variety of the malaria mosquito did not die out in Britain until after the First World War.

But malaria is not the only invader of temperate climes. Another possibility is an outbreak of bubonic plague. The brown rat and the flea which carry the disease have increased dramatically in Europe in the last five years in the mild winters. There were two substantial outbreaks of plague in the first quarter of this century in England in Suffolk and Bristol, so warming and an increase in rats and fleas makes another incident more likely.

Cases of plague have recently been diagnosed in Russia, North Africa and the United States. But while malaria and the plague make good headlines there are other less well-known diseases that are very unpleasant and kill or disable a number of people. Ticks, which are common parasites of cattle and sheep, survive far better in climates with less frost and the incidence of a tick-borne disease called Lyme disease is increasing in Britain. It is a disabling disease of the joints and an increasing number of deaths have been linked to the infection, particularly among country people and walkers.

Another disease spread by fleas that ride on rats and mice is murine typhus. This is associated with poor housing conditions and has been found in Mediterranean countries and as far north as New York. Scientists are surprised that it has not already reached suitable climates like the United Kingdom. One disease that has already been mentioned is leishmaniasis, caused by a parasitic protozoan transmitted by the bite of sandflies. The disease attacks all the body tissues and requires prolonged treatment. It affects about 12 million people a year and is fairly common in the south of France. One form of the disease is transmitted by sandflies of a type that now breed as far north as the Channel Isles. Scientists believe it will only be a short time before it is established in the tourist areas of the south coast of England.

Many existing pests are kept in check by the occasional sharp frost, perhaps the most obvious example in Europe is the greenfly. A series of mild winters have already allowed spring population explosions. Unwelcome insects and plants have been carried around the world by ships and more recently by aircraft. Normally they are rapidly killed by the unfavourable climatic conditions as they arrive. Some, however, survive and adapt to the new environment. Many tropical ants have found homes in centrally heated hospitals and blocks of flats. Scorpions that arrived on tankers and gained a toehold on Canvey Island in the Thames Estuary in England might soon find it is warm enough to expand their territory elsewhere.

But the health effects of global warming are not just about playing host to insects that used to live elsewhere and catching diseases or being poisoned because water is warmer than it used to be. A well-fed population is usually a healthy one and good nutrition is fundamental to health. Malnutrition is a major cause of infant mortality, physical and intellectual stunting in childhood and impairment of the immune system, thus opening people to infections. Currently about one tenth of the world's population is hungry and an even larger number malnourished, although the definition of what that means varies between scientists. The areas which will suffer most from lack of moisture have already been discussed.

However, there remains considerable argument about the total effect of global warming on agriculture. There is a strong body of opinion that the extra carbon dioxide fertilizer effect will increase yields. Some scientists say this is only temporary. In any event there are many other different regional effects. Some colder temperate regions make definite gains because it will be possible to grow wheat and other crops in areas where it was previously too cold. Large parts of the United States may lose the wheat harvest because it will become too warm and dry.

What universally worries scientists about health is the effect on populations in the tropics. In the previous chapters we have seen how regional temperatures may rise and rainfall reduce. The 1996 IPCC II report concludes:

## Population rise

Projected world population growth in billions

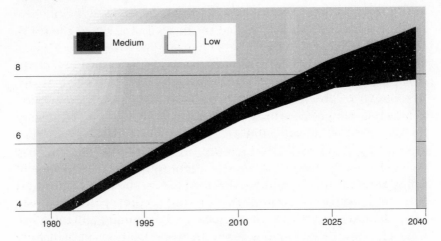

Since climate change may threaten food security in poorer countries within the semi-arid and humid tropics, poorer countries, already struggling with large and growing populations and marginal climatic conditions, would be particularly vulnerable to food shortages and malnutrition. In such countries there is minimal capacity for adaptive change.

There are already 100 million people in Africa who are food insecure. The cost of food on the world market is expected to increase if production declines in the 'mid-latitude bread baskets' like the United States, as it is predicted it will.

The large minority of the world population that already suffers from malnutrition would then face an increased threat to health from agricultural failure and rising food costs. A recent analysis predicts an extra 40–300 million people at risk of hunger in the year 2060 because of the impact of climate change, on top of a predicted 640 million people already at risk of hunger by that date in the absence of climate change.

An area barely touched on elsewhere is the effect on animal health. As with humans, there are a large number of parasites that affect animals; these too have a restricted range due to temperature, but are likely to spread. There is also evidence that farm animals, particularly young ones, suffer from heat stress, which in some cases is fatal.

The threat to fisheries because of the loss of the breeding habitat of the coastal ecosystem has been mentioned in previous chapters, along with the threat to human health in coastal communities where fish is a vital source of protein. While human intervention in the form of massive overfishing is the main cause of depleted fish stocks, the changing climate may also be a factor. At present, because of the crisis caused by overfishing, it is impossible to separate out any existing or future effect climate change will have on the abundance of fish in the sea. Water temperature and food availability have always varied enormously even at the same latitude because of water depth and currents. If climate change induces movements in these currents it could shift the migration and feeding pattern of fish. For industrial countries with sophisticated techniques, locating fish stocks is not difficult but it is again the poorer nations with inshore fishing boats that will suffer if stocks disappear or move out of reach. Overfishing is likely to mask any effects of climate change in the short term but its potential should not be underestimated. Combined with other problems which beset people in the developing world it could make survival impossible. Mass migrations of people who find they can no longer win even a subsistence living from the soil and sea are increasingly likely in a warming world.

The health of trees was mentioned in the last chapter. They are vital for the well-being of people, especially the surviving forest tribes, who carry in their folklore so much knowledge about the properties of forest plants still unknown to industrialized technology. For much of the world trees are a vital source of fuel and in many poorer parts they are rapidly disappearing. Forests also have an important role in climate, being massive users and storers of water. As we have already discussed elsewhere, on mountains they have the effect of slowing down the release of rainwater and so reducing the flooding further downstream. Any reasonably large spread of trees creates its own microclimate and frequently produces its own rainclouds. Many millions of people live in forests and make their living from them. The well-being of human settlements in the forests is clearly dependent on the continued ability of the trees to thrive.

# 13

---

# Escape plans for
# planet earth

IF NOTHING IS DONE to halt climate change then the whole of humanity is going to notice a considerable difference. As can be seen from previous chapters, the potential of global warming to cause death and destruction of both people and the natural world far outweighs any advantages. Having established there is trouble in store there are two options: to try to stop the process or to adapt. So far, the effects of climate change are relatively minor compared with what is to come. The optimistic view of the future is that if we can send satellites into orbit and men to the moon we can fix climate change too. The good news is that the optimists are right: we have the technology and the know-how, and if we work on it we can make it even better. Sadly, there are strong forces pulling against fixing climate change. We discuss some of them in the next section on politics, which shows how little progress has been made despite the urgency of the problem. It takes political will to get things fixed and that seems to be lacking.

In this chapter some of the options that could help to solve the problem and some of their relative merits are discussed. Many of them are remarkably simple and effective, and we should be doing them anyway, even if climate change was not an issue. These are policies that save money, reduce pollution and give people a more comfortable lifestyle. They are 'no regrets' or 'win win' policies. We know what they are, how to achieve them, and that we would all benefit, yet mostly they still remain undone. This is where the pessimists get a hearing, and rightly so; it is hard to believe, on the evidence, that *Homo sapiens* is an intelligent life form.

But let us get back to the science before we get on to the fixes. As has already become clear, there is a time lag to climate change – the effects of changes in the atmospheric concentrations of gases only occur many years after the changes have happened. It used to be thought that there was a minimum 30-year time lag, but it now seems that the adverse effects will go on getting worse for much longer. It will take centuries for the climate to get back to what is known as equilibrium, in other words for the weather to be stable, the ice caps to stop melting and the oceans to settle at a steady level. That is almost certainly true even if we stopped producing extra carbon dioxide and methane at this moment, but for every extra tonne of greenhouse gases that gets into the atmosphere the process of reaching a new level of stable climate is pushed back in time and the potential effects get worse.

Of course the greenhouse gases are also removed from the air by natural processes, some quicker than others, which is why scientists are most worried about carbon dioxide because it lasts longer in the atmosphere. The removal process means it is possible to reduce or even reverse global warming, even if it is over a long timescale. This could be achieved by cutting the world's emissions to a tiny proportion of the current figure and thinking of extra ways of soaking up the surplus greenhouse gases in the atmosphere.

Currently, all that belongs to the realm of scientists and dreamers who have never met any of the current generation of politicians, and have not realized that the men and women who represent us have feet of clay. Whatever happens it appears there must be some adaptation. The climate change we have already induced, which has not yet caught up with us, probably means that it is too late to prevent many low-lying islands and coasts losing some of their land. Cities, and other places with big investments like harbours, nuclear power stations on the coast, and assets like quality agricultural land will need to start adding to existing barriers to keep out the sea. Those who have not got the money for these defences or who realize it is impracticable or impossible to keep out the rising waters need to make alternative arrangements. From the biblical story of Noah onwards, and there are other better documented instances in past centuries when migration has occurred, rising waters have meant building and launching boats and sailing off to a new land. However, in a modern world where nearly every scrap of land is spoken for, or claimed by a nation which does not want new and poor immigrants, this will mean government intervention to find somewhere for the dispossessed to migrate to. This is also true of those who face starvation or other land damage as a result of climate change, and will be forced to tramp across continents to find a new home.

Politicians in Europe and elsewhere are already very sensitive and resistant to the idea of economic, or in many cases environmental, refugees. The scientific and economic answer to this is to build up the economies of the poorest countries so that there is an incentive to stay and in times of shortages they have enough economic clout to buy the necessary food to survive. This is an enormous issue which has been virtually ignored by all industrialized countries, who are currently cutting their overseas aid budgets. In other words, this issue has not been faced politically, and there is no sign that it is going to be. Meanwhile the pressure for mass migrations can only get worse.

But perhaps we should cheer ourselves up here with some simple ideas that work. Anyone who has been to the Mediterranean or seen pictures of villages in this region will know that many of the buildings are painted white. These people know a thing or two about keeping cool, and scientists, having first checked, recommend that as many buildings and roofs as possible across the world should be painted white or fitted with shiny surfaces. This reflects the sunlight back and reduces the 'heat island' effect which kills so many people in cities. Large conurbations are frequently 3 or 4°C (37–39°F) warmer than the surrounding countryside. In the summer this is caused by the heat-absorbing quality of tarmac, concrete and other building materials. Paint these surfaces white and it has the same effect as the snow – a large part of the heat is reflected back into the atmosphere.

Trees have the same effect of reducing the heat and incidentally the pollution, and if there are enough of them they will also fix a lot of carbon dioxide. One word of caution from scientists is that in clay soils trees could take too much moisture out of the soil and damage the foundations of buildings. Massive tree-planting schemes all over the planet have enormous potential in the next 50 years to soak up surplus carbon dioxide in the atmosphere. This would allow some breathing space while some of the technology mentioned below is developed.

But back to the main item on the agenda, which is how to reduce the artificial greenhouse gas emissions. For almost two centuries global energy demand has risen by an average of 2 per cent a year. The three sectors that are most important in producing carbon dioxide are industry, with 43 per cent of emissions, residential and commercial buildings, 28 per cent and transport, 22 per cent. These are 1990 figures, and already transport had been the most rapidly growing emitter of greenhouse gas of the previous 20 years and has continued to be so since. Without policy intervention all three sectors will continue to grow. We will deal with all three but first we come to the 'win win' options for all three. The main one is energy efficiency, with 10–30 per cent reductions in existing energy use and therefore carbon dioxide emissions possible with little or no net cost. Already several million jobs worldwide have been created in industrialized countries through the design, manufacture, sale, installation and maintenance of products which cut down on the energy waste and avoid carbon dioxide being emitted.

Examples of technologies already available include more efficient lighting, motors, heating and ventilation systems, office equipment and domestic appliances. In energy-efficient buildings, energy use can be reduced by around 80 per cent. Higher initial purchase prices of items are paid back in a matter of months or a few years. A report in the Netherlands, a country that is ahead of the rest of the world on these issues for understandable reasons, showed that carbon dioxide emission reductions of 80 per cent could be made over the period to 2050, while maintaining an annual average national growth rate of 2 per cent. Sadly, so far, many countries have done very little on energy efficiency, and all of them could do more.

Across the entire industrialized world more jobs could be created and cost savings made in the construction, manufacture and service industries. These improvements have the advantage of increasing the competitiveness of both countries and companies. Measures which boost energy efficiency are more labour-intensive and cost less capital than providing extra energy supplies and so have the twin benefit of reducing unemployment and releasing capital to be spent on other more worthwhile projects.

In former West Germany the production of energy-saving items, like insulation, increased 4.6 per cent per year in the decade to 1982.

By comparison, total industrial production increased by 2.6 per cent over the same period. The resulting direct employment increase in German energy-saving industries was 40,000 jobs, with another similar number in positions associated with planning, installation and maintenance. Overall, energy savings in former West Germany between 1973 and 1990 led to around 400,000 new jobs. (In 1973 OPEC countries gave the world a shock by using their monopoly to impose large price rises and incidentally at the same time gave a boost to energy efficiency.) In the United States the Massachusetts Energy Efficiency Council, a business association covering one US state, reported at the end of 1992 that its 750 member firms alone had already created 15,000–20,000 new jobs. Leading firms had quadrupled in size in four years. In six New England states overall, the energy-efficiency programmes are expected to create 30,000 new jobs during the 1990s. It will save around $6.7 billion in the period on electricity bills, and considerably reduce the greenhouse gas emissions.

Although forward-looking governments, companies and home-owners would introduce energy-saving improvements to save money, it seems from experience that a political push and further financial incentives are also required. This appears to apply across the whole greenhouse gas mitigation process and some of the incentives being discussed are detailed in a moment. Another key area where the need for political action also applies is energy production. The efficiency of converting fossil fuel to energy can be increased from an average of about 30 per cent to 60 per cent with existing technologies, and further gains are known to be possible with further development.

Some of the knowledge of how to do this has existed for years but is still hardly used. For example, the use of combined heat and power, where hot water generated in power plants is utilized for district heating schemes rather than wasted in the atmosphere, is only adopted in a tiny proportion of cases, although it is clearly very efficient.

Switching from coal to oil or natural gas, and from oil to natural gas can reduce emissions. Natural gas has the lowest carbon dioxide emissions per unit of energy of all fossil fuels, about two thirds of that of oil and almost half that of coal. Large and often untapped resources of natural gas exist in many areas. New low-cost, highly efficient combined cycle technology gas-driven power plants have reduced electricity costs considerably in some areas, as well as reducing emissions. Natural gas could potentially replace oil for some cars and lorries, particularly where it is cheaper.

From an environmental point of view one of the saddest sights, apart from an oil spill, is the massive waste of energy in oil fields. (On a visit to Iran after the Gulf War to study the environmental damage of the oil well fires I was struck by the hundreds of gas flares all over the Iran oil fields. Millions and millions of tonnes of gas was being burned without any attempt to use it. The technology to capture the gas and use it for industrial purposes exists but is denied to Iran for political reasons. The fault lies on both sides for this continuing waste of resources which damages us all.) A technology with potential which is still in the experimental stage is taking the carbon out of fossil fuels to make hydrogen-rich alternatives – as yet this is far from economically viable. Leaving this and other experimental systems aside, even with existing technology, efficiency gains of 50–60 per cent are possible from the current generation of power stations, according to the scientists. Even if this was achieved it would still not be enough because the worldwide energy demand is still growing and emissions in absolute terms would still rise.

The most obvious greenhouse gas reduction method is building up the already existing non-fossil fuel sources of energy: nuclear power and the vast array of what are called 'renewables'. Nuclear power was once the favoured option of the industrialized world but it has fallen foul of public suspicion on safety grounds. Leaving those aside, the economics of nuclear power remain unconvincing, especially where the unresolved problems of disposing of nuclear waste are concerned. Incidentally the uncertainties of climate change make safe disposal even more difficult.

Renewables are potentially capable of providing almost all the world's energy needs. Putting aside burning wood, which is the oldest form of heating, hydro-power has been the most readily exploited so far. In long-standing schemes, once the capital cost has been written off, it is by far the cheapest source of electricity. In parts of Iceland where schemes have been in place more than 30 years and the turbines are virtually maintenance-free, the charge for electricity would be nil apart from the need to build up capital for new schemes to meet greater demand. Hydro-power has its own environmental disadvantages and big dam projects are frequently heavily resisted by the people who are turned off their land. Most large-scale dam developments have turned out to be economic failures with the optimistic economic forecasts unfulfilled.

As we have seen from previous chapters, dams tend to cause problems downstream by preventing the flow of silt. Despite the dismal track record of these mega-projects the number of large dams being built worldwide continues to increase and they are heavily subsidized by the World Bank and other aid institutions. Smaller hydro-projects, however, which are far more beneficial, are seriously unexploited. They have great potential in any part of the world with reasonable rainfall. Turbines generating electricity can be placed in most rivers in series without any environmental damage, repeatedly using the same water to provide power for local needs.

The other renewable technologies are still in their infancy but in 1990, taking all renewable sources, they made up 20 per cent of the world's primary energy production. Geothermal power is one of the better established and has great potential. In parts of the world where there is volcanic activity or the earth's crust is thin it is possible to tap into the heat and use it to generate electricity on the surface, or directly heat water for domestic or industrial use. Iceland and Italy are pioneers in this area, but other countries with less obvious volcanic sources are also experimenting using modern technologies to extract the heat from deep underground.

Among the newer technologies for producing electricity, although one of the oldest for grinding corn, is wind power. It is now an established industry in Europe and the United States, and has enormous potential along with less developed but promising experiments with wave power. But back to the oldest form of power for a moment – burning wood. This idea began to find political favour in industrialized countries in the 1990s because of its potential to make use of derelict or surplus land for fast-growing trees. These provide a continuous harvest for wood-burning power stations. These power stations are already in use, burning what is called biomass, which can be anything from straw from wheatfields to sawmill waste, and have great potential as a renewable energy resource, providing badly needed new employment in country areas.

The greatest single potential for the future is solar power. It is clear from the science section of this book that the most powerful source of energy available to us is the enormous power of the sun, which keeps us all warm. Large areas of the earth have hot sunshine

140

for most of the year, and even in temperate zones there is enough to make a substantial contribution to heating and hot water.

Already small calculators and watches are driven by sunlight and many countries have solar panels for water-heating. The real breakthrough will come when the power of sunlight can be used for continuous electricity generation. Perhaps the best hope at the moment comes from photovoltaic solar cells. These use a thin slice of silicon to turn solar power into electricity, but although they work they are not yet fully competitive in price. Greenpeace, among other organizations, are convinced that the solar revolution is the way forward and serious money is being spent by industrialists who see fossil fuels as yesterday's industry. By the turn of the century the photovoltaic power cell is expected to be big business.

Trying to put these rising technologies in context with how the world currently generates power is rather difficult. Fossil fuels still dominate and in the developing world, particularly China, which has vast coal reserves, virtually all new power stations currently being opened are of the dirtiest, less efficient variety. One important factor to remember, however, is that even in the most backward countries the world's commercial energy system will be replaced at least twice in the twenty-first century. That means even without premature retirement of existing stock, significant improvements can be built into every replacement. Indeed for every power station that is shut down through old age it must be sensible for every company to review its cheapest option for replacement. There are what the scientists and economists call 'institutional problems' here. These take a variety of forms but many of them are political in nature.

For countries with large coal reserves and a large workforce of miners, it is very difficult to abandon the idea of building new coal-powered stations, and to substitute solar power panels imported from abroad. Similarly, in Britain the electricity industry can only make profits by selling more electricity to its customers so it has no interest in selling energy-efficient equipment. However, it is possible to encourage energy efficiency through government policy. In the United States, power companies were given an incentive to go for least-cost planning, in other words to decide whether it was cheaper to build new plant or provide their customers with the goods, equipment and advice necessary to reduce electricity demand. They were allowed to make at least as large profits from making homes, offices and factories more energy efficient. It proved cheaper in every instance, at least in the first stages, to insulate homes and take other efficiency measures than to build new power plants. The next stage is to consider the most energy-efficient form of generation. These kinds of simple alterations to the way the energy market works could certainly operate in Britain and other European countries, but although the governments have repeatedly been made aware of them, few have moved to follow the American example. But more of that in the politics section.

One of the problems the economists and scientists have tackled and not come to a satisfactory solution on is costing environmental damage. It is clear that the rich countries in their development have created the climate crisis we are in, yet it is the poor in blameless countries that suffer most of the effects. In their third 1996 report, IPCC III, the scientists put it this way:

> It therefore becomes a highly political question of how much the world cares about imposing risks upon some of the poorest and most vulnerable people.
>
> If the success of international responses is measured in terms of aggregate impacts on global wealth, then their position will not feature. If it is judged by how well it protects the weakest, and minimizes the most severe suffering an entirely different approach may be called for. Thus it needs to be recognized that attempts to quantify the costs associated with climate change involve inherently difficult and contentious value judgements, concerning which different assumptions may greatly alter the conclusions.

In practical terms, they are saying that if emissions from coal-fired plants in the United States and Europe contribute towards the disappearance of the Maldives a price per tonne of carbon dioxide could be charged by way of a penalty. The money could then be used in solving the problem. In this same argument there has been a row about the economic value of human life. From an economist's point of view, a life in the industrialized world appears to be worth more than one in the developing world, but more about that later. Accepting that the environment has some value, and that the polluter must pay something towards the damage caused, even if it is an entirely arbitrary figure, opens the door to what is known as a carbon tax.

Currently some fossil fuels are subsidized by governments so removal of all subsidies would have to be the first economic decision, immediately followed by gradually increasing taxes per tonne of carbon dioxide released as a penalty for the contribution to climate change. If fuels were charged strictly for each tonne of carbon dioxide they release into the atmosphere, coal would suffer most, and gas least. Renewables, which would not be charged at all, would be made more competitive. There are suggestions that the revenue raised from these taxes should not go into general funds but should be redirected towards research and development for renewables and paying for energy efficiency schemes. This would create a large number of new jobs.

In the industrial sector, leaving aside power generation, there is expected to be a cut in greenhouse gas emissions because of efficiency savings. About 30 per cent savings are possible without any net cost – another 'no regrets' policy waiting to happen. The potential saving is much larger with imaginative use of energy and recycling of materials which would otherwise be wasted. Simplified methods of production are proving strikingly successful in pilot projects and companies have saved millions of pounds a year.

Transport is the big headache, not just for greenhouse emissions. The air pollution problem in many cities is already life-threatening without any need to worry about its further effect on climate change. Curing this is in itself a priority and any gains for reducing global warming make this another 'win win' area.

Scientists estimate that by 2025 emissions can be cut by about a third with more efficient engines, lightweight construction and improved design, without any drop in comfort and performance. But that is clearly not good enough, not least because of the massive increase forecast in private car ownership. Governments will have to force the pace by adopting a series of policy measures. They must put on pressure for smaller vehicles and improve public transport making it cheaper than private car use, but that is only a start. The trend to long-distance commuting, shopping out of town, and more and more freight by road rather than river and train must be changed too.

Because of the difficulty of reducing carbon dioxide emissions, and the fact it is the most important greenhouse gas, little attention has been paid in these pages to the reduction of methane. The scientists take the same view, that by comparison methane is a minor worry, partly because it disappears faster from the atmosphere. Most methane is produced as a result of agriculture and, as has been said elsewhere, attempts to develop better ways of growing rice are being made. Going back a long way to the flatulence of termites and farm animals the scientists are suggesting that farmers should put their ruminants on a diet which minimizes some of their production of wind. Better management of manure and human waste tips are all suggested. Extra methane emissions could be cut by up to 50 per cent by these methods.

The scientists and economists have constructed computer models to test out all their theories; after all 700 possible policies and measures have been put forward to reduce greenhouse emissions. They call them 'thought experiments' and they factor in such things as global population increases, and economic growth. They optimistically predict that they could reduce global carbon dioxide emissions from 6 gigatonnes in 1990 to about 4 gigatonnes by 2050 and 2 gigatonnes by 2100. That, they reckon, would work in theory if there was enough political will and considerable investment in alternative energies. It is good to know we could do it. In the next section we examine what the politicians are doing with the opportunity offered.

# Part Four

———

# THE POLITICS

# 14

---

# Berlin:
# the climate debate
# comes alive again

WHILE THE SCIENTISTS were updating the work of discovery and prediction discussed in the last two sections, what were the politicians doing? Almost three years had elapsed between the Rio Earth Summit in 1992 at which the Climate Change Convention was signed and Berlin's First Conference of the Parties (COP1) which began on 28 March and lasted until 7 April 1995. Although the scientific evidence was becoming more alarming all the time, the issue had gone to sleep as far as the general public was concerned. There is an argument often used that issues like global warming go in and out of fashion, that newspapers and television need to pass on to something new. There is some truth in this but to stay on the news agenda a subject has to have something fresh happening all the time. So despite appearances to the contrary, many issues do not go away, they just wait to be rediscovered when a new fact or trend emerges. In this case, as we have considered in the last two sections, the scientists had continued on their way. To some extent so had the politicians, but as we stated earlier, once a treaty has been opened for signature there is always a delay before individual governments ratify it. Then a further grand conference has to be convened to decide what to do next.

At the Earth Summit the Germans, who were particularly keen to take action on global warming, had agreed to host that first convention meeting in Berlin. But before we discuss the events at COP1, it is important to put it in the correct context with the science. By concentrating in the two preceding sections on the most up-to-date information available we have allowed the science to jump ahead of the political narrative. Although an interim IPCC report was prepared in time for the COP1 meeting, the science we have been reviewing was not brought fully up to date until a year after the Berlin conference. The crucial point about that was that politicians who wanted to do nothing were still able to emphasize that uncertainty still surrounded the science. The scientists had still not come out emphatically, as they have since, to say that climate change due to human activity was underway. This bold change has made an important political difference which we shall consider later.

There were other factors which had taken the issue out of the headlines. The worldwide recession, which had already begun by the time the political leaders met in Rio, had continued. There it had caused an extra bout of meanness among the industrialized nations asked to help the poorer nations cope with the consequences of climate change and other environmental degradation. Since then the concerns of ordinary people had also switched away from the environment. Ordinary voters were more worried about keeping their jobs than saving the world from global warming.

In fact many trade unionists and politicians from the right and left, who were ignorant of the real arguments or happy to distort them, often portrayed care for the environment as threatening jobs. Mount Pinatubo's eruption, so important in one sense for the scientists to prove their theories that global warming was happening, had also caused temperatures to cool. It meant that between the two key political conferences there had been no juicy headlines and opportunities for journalists to go on about yet another hottest year on record.

So the environment had slipped down the political and news agenda for perfectly understandable reasons, but this does not mean that nothing had been happening. The sherpas, as civil servants who do all the work on international negotiations often refer to themselves, had been to many meetings to try to take the issue forward. Under the Climate Change Convention a body ponderously called the Intergovernmental Negotiating Committee (INC) had met no less than 11 times in an attempt to prepare for the Berlin meeting. The INC had explored every possibility but had eventually reached a stalemate. The last meeting of the INC in New York in February, just before the Berlin meeting, was supposed to pull a rabbit out of a hat. The civil servants had been allowed to agree on behalf of their governments that commitments made by the industrialized countries to reduce emissions were inadequate. This suggested that a new agreement was needed if nations were to conform with the objects of the Climate Change Convention.

One radical proposal was on the table, unanimously supported by the environmental groups of the developed world. This was from the Alliance of Small Island States (AOSIS) and was referred to as the AOSIS Protocol. As these 36 countries from the southern Pacific, Indian Ocean and the Caribbean constantly reminded everyone, they would be the first to disappear if nothing was done. Their basic demand was a 20 per cent reduction in carbon dioxide emissions by the year 2005. This was the original target agreed seven years before at that optimistic meeting in Toronto when climate seemed a much easier problem to tackle. It was also known as the Toronto Target.

However, in 1995 in the midst of a recession with economic costs on their minds, the politicians from the United States and other big carbon dioxide producers were not prepared for any reductions in emissions. The sherpas must have arrived in Berlin depressed and despairing of the Convention's future.

But, as described in the section on the history of climate change, remarkable things happen when politicians are faced with a high-profile conference in which their personal reputations are at stake, particularly when there are a large number of reporters present. Not that Berlin had anything like the glamour of the Earth Summit; instead of a large number of high-profile signing ceremonies and big speeches, most of the conference was about trying to hammer out some way forward where previous meetings had failed. As stated in the earlier section when discussing how the Climate Change Convention came into being in the first place, there was very little by way of concrete commitments. Because of the Bush administration's stance in 1992, countries had only signed up to promises to get carbon dioxide emissions in the year 2000 back to 1990 levels.

It was not a legally binding agreement, although it would be embarrassing if any of the big countries failed. The Bush administration had been replaced by President Clinton and his allegedly very environmentally aware vice-president Al Gore. But already in 1995 it was clear that the right-wing Republican backlash was on the way. These Republicans wanted less government, less regulation, less interference in an unfettered free market. They thought the environment needed less protection, not more. In this context the second big unresolved question on what to do about emissions after the year 2000 was a difficult one to answer.

Critics of the original treaty were keen to point out that there was, for example, nothing to stop any country dramatically increasing its emissions after the turn of the century. But they were neglecting the strength of the Convention. This was the statement of the clear objective 'to prevent dangerous man-made interference with the climate system' and then to put the onus on those who had ratified the agreement to adopt policies and measures to achieve a safe climate. One of the issues not yet touched on but still to be clarified was what constituted 'dangerous interference' in the climate. It meant that notionally, at least, there had to be some ceiling fixed for a maximum concentration of carbon dioxide and other greenhouse gases in the atmosphere. This accepted that emissions would rise but only up to a certain level at which point the human race and the rest of the ecological system could adapt to a warmer world, provided it did not all happen too fast. Calculating what this limit should be had not even been properly addressed before Berlin, but at least the issue had begun to be discussed.

In other words, those gathered at Berlin had come to address some difficult decisions and find a way forward beyond the year 2000. In theory the treaty could be amended but the most likely course was the creation of a new legal agreement called a protocol. In order to work, this had to be devised and agreed by all parties. However, this was what 11 preparatory meetings had conspicuously failed to do.

There were other unresolved issues. Technology transfer from the industrialized world to the developing world was envisaged as part of the original Convention, but how was it to be organized and who was to pay? How would the companies who developed the technologies benefit if they had to hand over their work free of charge? Technical advantage usually meant economic advantage, so technology transfer involved loss of what they called intellectual property rights. This boiled down to loss of profit as a result of handing over the nuggets of research – not an easy idea to sell to the energy industries.

At the Earth Summit there had been a dispute over how the funds from the industrial world were to reach the developing world. The World Bank might seem the obvious candidate but its environment record was shocking and its secretive decision-making process meant that it was regarded with great suspicion by Third World countries and the environmental movement. But the United States wanted it to handle the financial side of the Climate Change Convention. In 1990, when environment concerns on many badly handled aid issues had embarrassed the World Bank, a hastily cobbled together Global Environmental Facility (GEF) had been organized. This was to be operated by the World Bank, the United Nations Development Programme and the United Nations Environment Programme. It was a compromise to answer some of the criticisms of the lack of environmental care and control by the economists at the Bank.

In 1994 the GEF was adopted as the 'interim financial mechanism' for the Convention but at the time of Berlin less than £2 billion had been pledged to the fund for the period up to 1997, only 40 per cent of which was for climate-related activities. As the coalition of environment groups known as the Climate Action Network remarked in its briefing paper for Berlin: 'The World Bank alone will invest ten times that figure in largely climate unfriendly energy and transport projects over the same period.'

Another thorny issue, which was destined not to be resolved at Berlin, and may not be until the twenty-first century, was the choice of a system of voting to allow conference decisions to go through. This may sound crazy because otherwise how is the Convention to work at all, but how to make decisions stick is fundamental to the working of international conventions. Climate change talks work on the basis of consensus, in other words everyone has to agree to the final text of any document or decision before it can be adopted. This can be, and frequently has been, incredibly time-wasting because at any point any country can raise an objection. It has been used time and time again by the oil and coal lobbies to weaken the wording of both scientific and political documents.

By objecting to scientific certainties and building in 'maybes', 'perhapses', 'probablys' or even better 'possiblys', the impact of key documents has been softened. What many, particularly the more environmentally active nations wanted was an adoption of majority voting, either three quarters or four fifths of the countries being able to outvote the rest. The problem is that those who suspected they might continually be outvoted by this process were unlikely to agree to it so a consensus on a new voting procedure seemed impossible

150

to reach. Nevertheless, there are problems with majority voting in such a forum. It is more difficult afterwards to get countries who voted against measures to comply, a vital ingredient in climate change since it entirely depends on the world acting together. If, on the other hand, these dissenting nations accepted a consensus, even a watered-down version, then morally they are in an entirely different position.

In Berlin no resolution was found to this issue and as already stated, at the time of writing it is still being passed on as an uncracked nut from meeting to meeting. Nevertheless, unwritten rules began to develop at Berlin. Mrs Angela Merkel, an East German, who had taken over the mantle as Germany's environment minister and was chairwoman of the conference, exploited the consensus idea to her advantage. An unknown quantity at the start of the political proceedings, she turned out to be tough, energetic and astute. As all-night sessions developed among key countries seeking for deals to save the conference she was ever-present, but most important when the issues came back to the conference floor she made sure nothing stood in the way of a stitched-up deal.

It would have taken a brave Saudi Arabian delegate or anyone else to dissent in front of Mrs Merkel and the assembled delegates of 150 countries. When she banged down her gavel at the close of the conference she had sealed a consensus deal, far more effective in the end than a majority vote.

COP1 began on 28 March and involved thousands of sherpas and a mass of non-governmental organizations and delegations of a size not seen since the Earth Summit. These were now definitely divided into two highly organized groups: the environmental lobby which wanted action on global warming immediately, and the industry lobby, which wanted delay as long as possible. The struggle for supremacy between these two groups will be examined later, but the influence of both sides at Berlin and, more crucially, in the future will make all the difference to political progress.

Agreement in the first week of the conference had been as limited as at the previous talks, so when the substantial frame of Dr Helmut Kohl, chancellor of the Federal Republic of Germany, took the stage in the giant conference hall in Berlin there was talk of a complete failure.

The first paragraph of the *Guardian* story on the morning he was to open the political part of the conference read: 'The world climate conference in Berlin was in crisis last night as ministers arrived from around the world to find none of the substantive issues had been resolved, despite ten days of intensive negotiations.' Under the headline 'DEADLOCK OVER GREENHOUSE GASES' the *Daily Telegraph* spoke of Chancellor Kohl's embarrassment of opening the ministerial session with no agreement between industrialized countries to reduce or limit fossil fuel emissions after the year 2000. *The Times* went further: 'CLIMATE SUMMIT DISINTEGRATES AS NATIONS SQUABBLE'.

Chancellor Kohl's speech was to begin a series of political statements in which various countries, in complete contrast to the months of deadlocked meetings, were to demand action where none was currently forthcoming. Journalists had turned up in force for the first time on the day the politicians arrived. They, and their editors, were ready to write off world co-operation on climate change as an impossible dream. In order to give some flavour of the sudden change that an influx of politicians closely watched by journalists can make to proceedings there follow some selected extracts from the political speeches given on that day. Some of the climate change issues already touched on and some new ones are covered in these speeches. What was actually said on the day is reported.

Chancellor Kohl is given the lion's share of the space because he spoke first to a deadlocked conference. He was clearly prepared to put his political reputation on the line to make the negotiation a success even at the eleventh hour. Many others who spoke in his wake echoed his sentiments but he set the tone and in doing so set out his own minimum requirements.

For me the Rio Conference remains a mandate and an obligation. The countries of the world met in Rio in 1992 to face the fundamental task inherent in environment and development. On that occasion the world took up this matter as a central theme of international politics. It documented its readiness to go beyond very different starting positions and interests to look for joint approaches towards solving the problems. Rio provided a decisive impulse, not to say a clear signal of hope.

As a consequence of the worldwide recession of the last few years, however, this promising process has not developed the momentum we expected then. National egoism came more to the fore. The goal was economic recovery, the environmental

152

requirements were often neglected. Joint endeavours towards forward-looking developments that would secure the future were placed on the back-burner as allegedly expensive luxuries. This shows that the concept of sustainable development, the key message from Rio, does not yet sufficiently determine the thoughts and actions of states.

However, it is a dangerous mistake to believe that positive economic development can be achieved in the long term at the expense of the environment and of nature. We must not ignore the fact that global environment problems are continuing to increase rapidly. No country on earth is in a position, now or in the future, to overcome on its own the dangers arising for itself and its people from global climate change. What is needed, therefore, is joint action.

The greenhouse effect caused by us human beings thus threatens to thwart our efforts towards economic development and increased prosperity. The climate-related natural disasters seen in recent years have caused substantial damage to economies worldwide. The Alliance of Small Island States points out with some urgency that a further increase in global carbon dioxide emissions threatens their very existence, this is the survival of their countries and of their people. In order to secure the survival of humanity it is necessary to harmonize economic, social and ecological development.

Preserving Creation and securing sustainable development is a task for the whole of mankind – a task which no one – no state, no economy and no individual – must shirk. The number of people on our planet is growing rapidly. As a result the burdens on the earth's ecosystems are growing too. All the more reason for us to stop the reckless exploitation of nature in order to preserve development opportunities for succeeding generations.

He went on to detail the threats to the environment, the ozone hole, overfishing, desertification, pollution of water, and loss of forests, and then returned to the central argument of the conference.

Ecology and the economy are compatible. To this end we must look for intelligent solutions which link environmental protection and economic development. By improving energy efficiency – for instance, by building modern, more efficient power stations – we can not only supply more people with heat and electricity using the same quantity of coal or natural gas, but we can at the same time reduce the resulting environmental pollution.

Modern technology not only makes a large contribution towards ensuring that our environmental conditions allow a dignified life in future, it also makes economic sense in the long term.

The Chancellor then dealt with the effects on Germany's greenhouse gas emissions of the reunification of East and West. As a result of the exposure of East Germany's industries to competition there had been mass shutdowns. This had reduced the carbon dioxide emissions of the former East Germany by 43 per cent between 1990 and 1994. The aim for the whole of Germany was to break the link between growth and pollution. The target of reducing by 25 per cent the whole of Gemany's emissions of carbon dioxide from 1990 levels by the year 2005 remained while at the same time the country intended to continue economic growth. The Chancellor continued:

In Rio we agreed to reduce the production of greenhouse gases to 1990 levels by the year 2000. We must, however, ensure that emissions do not start to increase again after that date. The development over the past few years has shown that it is economically and technically possible to adjust to ecological necessities if the political will is there. As a first step towards effective climate protection, I call upon all industrialized states to follow the example of the European Union and to join us in the commitment to stabilize the emission of carbon dioxide beyond the year 2000. The necessary decisions must not be put on ice.

Above all, the Berlin conference must herald a further limitation and reduction of climate-damaging greenhouse gases after the year 2000. Therefore I appeal to all those taking part in this conference to adopt a substantial negotiation mandate here in Berlin with clear terms of reference aimed at a protocol that is binding in international law.

He went on to say that the new agreement should cover not just carbon dioxide but methane and all the other greenhouse gases.

It should also lay down clear targets for their reduction so that the world knew what it was aiming to achieve to stabilize global warming. It then needed to determine the necessary measures to achieve this aim.

Having dealt with the duties of the industrialized nations in reducing greenhouse gases, Chancellor Kohl went on to address the vexed question of finance, which had caused so much trouble in Rio.

> It is of no help if the industrialized countries make environmental demands which exceed the economic or financial resources of some developing nations. Finding common solutions always means sharing the burdens in a spirit of solidarity. It is immaterial for the climate of our earth which country or factory emitted the carbon dioxide or another climate-damaging gas.
>
> In our joint battle against these greenhouse gases we should therefore consider how we can achieve a substantial transfer of knowledge and technology to the developing countries and at the same time use the funds globally available for climate protection as effectively as possible. One promising way of combining both goals is the 'joint implementation of measures' already envisaged in the Convention.

This reference to joint implementation, or JI as it became known, was one of the most controversial in the speech and one of the great areas of disagreement at Berlin. It is one of those arguments that continues to rage in the corridors of conferences on climate. Currently it is in a 'pilot phase' and whether the idea would work, and be politically acceptable, is discussed later in this section.

Basically, the idea of JI is that industrial states may be able to make a greater impact on climate change by forming partnerships with other countries and taking measures outside their own borders to mitigate emissions. For example, if Germany spent money in Poland building new power stations it might reduce Europe's total production of carbon dioxide by a larger amount than if it updated its own already high-quality stations. The donor country would then get 'credits' for reducing world carbon dioxide emissions and be exempt from reducing its own levels by that amount. In practice, JI was highly controversial because developing countries saw it as a get-out clause for industrial nations who would thereby avoid having to tackle their own emissions.

Back on safer ground, Chancellor Kohl said climate protection was a joint task aimed at safeguarding the future. A habitable environment was something to which all human beings had a right.

> No reasonable person eats all the seeds of his crops because by doing so he is thwarting the future harvest. We must not destroy the soil on which our children's future must grow. Not only do we have responsibility towards those people who need food, work and social security today, but also towards coming generations.
>     Neither short-term thinking nor shying away from uncomfortable decisions must determine our course of action. We should not be content with noncommittal declarations, but achieve a breakthrough now on three central issues:
> 1. The industrialized countries have the responsibility to limit emissions permanently beyond the year 2000. This is a vital first step. We should stand by this goal.
> 2. With a substantial negotiation mandate we must set the course here in Berlin for a noticeable reduction in climate-damaging greenhouse gases after the year 2000.
> 3. Industrialized and developing countries should reach an agreement on the joint implementation of climate protection measures and thus make possible the necessary transfer of know-how and technology.

The Chancellor's bold approach was echoed by other European ministers. France, holding the presidency of the European Union at the time, was able to speak on behalf of all Europe in pledging to peg emissions at or below 1990 levels after the end of the century. John Gummer, the British environment secretary, was able to go further. The United Kingdom was already on target to reduce greenhouse gas emissions below the 1990 level by 2000. By the standards of previous British statements and those of non-European industrialized nations Mr Gummer's contribution was therefore revolutionary:

> An agreement on a figure in the range of 5 to 10 per cent reductions below 1990 levels by 2010 would seem to be a credible and achievable outcome of the negotiating process we are about to launch. . . . In dealing with this complex and difficult issue, we must not be trapped into making a false choice between our children's interests and our own shorter term interests.

Afterwards, speaking to reporters Mr Gummer said the United States had to recognize climate change as an issue that required action. The Conservative parties of Europe had embraced the concept of conservation – it was a lesson the right in America must learn. 'We are no longer talking about the effect on our grandchildren but on our children, and that means American children too.'

The United States undersecretary of state Timothy Wirth, like William Reilly at Rio, had the unenviable task of representing the Clinton administration. He made a bland statement which, according to my report for the *Guardian* at the time, 'skirted all but the vaguest mention of carbon dioxide cuts, speaking only of "continuing the trend of reduced emissions".'

The following day the tone of the coverage had changed. The *Daily Telegraph* had 'KOHL LEADS THE ATTACK ON US OVER POLLUTION' and the *Guardian* 'US CHALLENGED ON GLOBAL WARMING'. The *Financial Times* picked up another political undercurrent, which was the changing attitudes of the developing countries to global warming. Chancellor Kohl had already mentioned the plight of the island nations which could disappear if the sea level rose as predicted. As has already been mentioned, many other countries were also likely to suffer serious consequences from the same cause. This issue began to change the view of the developing world. Some 42 of the 'group of 77' had broken away and prepared what was known in conference jargon as a 'green' paper. These countries firmly sided with the AOSIS countries, the small island states who were already making emotional appeals on the conference floor to save them from drowning.

This marked a substantial change from the attitude adopted at Rio. The developing world there had regarded greenhouse gas emissions as the problem of what were called Annexe I countries, the Western-style democracies of the industrialized world and the old Communist bloc. After all, industrialized countries' emissions were as much as 20 times those of the developing world. But these countries now regarded the Rio commitments on global warming emissions to be 'inadequate'. They were acknowledging for the first time that it was a mutual problem, even if they were not accepting for the time being that they had any responsibility to reduce their own emissions. Indeed, until the industrialized world proved that it was serious about its own emissions they did not expect to be asked.

157

Angel Alcala, secretary of the environment of the Philippines, speaking on behalf of the group of 77 developing countries and China in the wake of Chancellor Kohl's call for action, had spelled out this significant change in approach from the countries of the South. The Philippines, the home of Mount Pinatubo, had been hit by a series of natural disasters. He began with a broad picture before indicating a split which was to become permanent between the majority of the developing world and the OPEC oil states:

> We are united on the conclusion that present commitments in the convention remain unimplemented. We also come to underscore the importance to developing countries of the transfer of technology, of financial resources, and the intimate links of environment and development.
>
> In the universal effort to conserve the environment, we have been guided by a belief in nations' right to development; by the principle of common but differentiated responsibilities, and by the assurance that our partners would consider the specific needs and special circumstances of developing country parties.

On the issue of the adequacy of commitments the group was united against any attempt 'to impose any new and additional obligations on developing countries'.

He then moved on to the issue of the green paper and revealed that from the original 42 signatories under the leadership of India there were now 70 countries in support, including China. They were backing the demand for 20 per cent reductions of carbon dioxide emissions by 2005, the most radical proposal before the conference, previously endorsed only by the AOSIS countries. 'The adoption of this target being a minimum condition to make this conference a success.'

Then, solely on behalf of the Philippines Mr Alcala continued:

> We will do our part in responding to climate change. We are conducting an inventory of our greenhouse gas emissions. We are committed to the sustainable management of our forests and the protection of our coral reefs, thus conserving and enhancing carbon sinks. We have enshrined energy efficiency as a cornerstone of our future energy systems. We are also completing a country action plan on climate change.

He talked of the need for a new standpoint,

> not of a neutral observer, not of indifference, but of courage, commitment and vision. Courage to change wasteful consumption lifestyles. Courage to move away from obsolete energy systems. Courage to agree to a Berlin Green Mandate. A courage based on commitment. Commitment to sustainable development. Commitment to justice and equity. A commitment founded on vision: of nations and states united in one common purpose; of people and communities caring and loving, each other and our planet.

Chancellor Kohl had begun by throwing down the gauntlet to the conference and in demanding action he had deliberately embarrassed the United States. To everyone's surprise the developing world had picked up the gauntlet and thrown it back, challenging everyone to pick it up. The sherpas must have groaned. This kind of talk meant that the politicians needed a deal to take home, and it would be the sherpas who sat up all night with instructions to get one. They must have wondered why they were not told all this months before when they had all the time in the world. Sadly politics is not like that. New and lasting alliances had been formed in the back rooms of the conference centre in Berlin and as a result in the next two days the issue of climate change moved forward quicker than it had done in the previous two years.

# 15

## Hope salvaged
## from the wreckage

EVEN AS THE RHETORIC demanding action was echoing around the conference hall, the sherpas were beavering away in closed rooms with new instructions. Outside the conference hall at one of his few press conferences in front of a hostile audience Timothy Wirth, the United States undersecretary of state for global affairs, told journalists:

> The US government is not going to commit itself to things it cannot do. We will be fortunate if we can keep this treaty alive. I suspect if we do we will still be talking about climate change and how to combat it in the years 2000, 2010 and 2020. We still have three days to go.

In the negotiating sessions there were now four political camps pulling in different directions but all anxious for a face-saving deal. The real negotiations were going on in 'room 7', another conference hall in the giant Berlin centre where entry was strictly controlled. It would be here, if there was to be a deal at all, that the result of the Berlin conference would be decided.

Before the conference had begun it was clear that the old North–South divisions on what to do about global warming had become far more complex, both camps having subdivided. Berlin crystallized these divisions and, as has already been seen with the 'green paper' from the developing world, each group produced its own negotiating position.

Among the industrialized nations it was the United States which got all the headlines and most of the stick for slowing the process towards action. By comparison the Europeans, and particularly the Germans and Nordic countries, except Norway, were seen as the progressive camp, with targets and timetables. The former Soviet Union and the new states spawned from it were suffering such industrial chaos that they were not sufficiently organized to make much impact either at Berlin or since. The emissions from these states had gone down enormously by default as industries collapsed in the wake of Communism, but targets and timetables are meaningless when a struggle for economic survival is the main concern. These nations are still included in the process, however, and are expected to play their part. This is particularly so in the energy sector where decisions being taken in the late 1990s will affect eastern Europe's emissions for the next 30 years.

But back to the more predictable countries, at least in the economic sense. In reality the United States was part of a substantial group of industrialized nations known by the cumbersome title of the JUSCANZ – the jumble of letters representing the countries who informally supported each other. To spell them out, the main players who gave their names to the group were Japan, United States, Canada, Australia and New Zealand. Extra supporters of JUSCANZ at Berlin were said to be Switzerland, Norway (with its oil interests), Mexico and Iceland. Leaving aside Japan, all the main players had high emissions per person and very little prospect of getting the total down in time for 2000.

Canada, with a population sold on energy-profligate lifestyles similar to those of the United States, lived in its neighbour's shadow at the climate talks, and largely escaped the flak as a result. Mexico had other reasons for supporting the United States, mainly cash and free trade considerations. Australia and New Zealand, on the other hand, had gone in a short time from being campaigners for big cuts in emissions to being nervous about reaching any targets at all. Australia attracted attention from environment groups which saw the government's behaviour as treachery. Australia now presented itself as a 'special case'. Having originally supported the Toronto Target of 20 per cent cuts by 2005, Australia looked likely to overshoot the stabilization at 1990 levels by the year 2000.

In fact, Australia's per capita emissions were among the highest in the world. The government was claiming that this was caused by the high energy requirements of its industry, particularly the production of metals. There is some validity in this argument but there were large parts of Australian industry, for example chemicals, which required little power. According to environment groups this balanced out the demands of other energy-hungry industries. A political factor was the extremely strong coal lobby in Australia, cheap coal being an important export. Politicians caught between the green lobby and industry had sided with industry – but more on that in the next chapter on the power of lobby groups. One of the main points about Australia's difficulties was that energy efficiency in Australia was poor but politically difficult to combat, particularly during a serious recession.

This left Japan, which had altogether different reasons for not wanting to reduce emissions. Being a country without oil it has always been conscious of its vulnerability due to its lack of indigenous energy. This is why, despite so much hostility to all things atomic, the government had sanctioned a nuclear power industry to try to make the country less vulnerable to being cut off from fossil fuel imports. As a result of this in-built nervousness the 1973 oil price rise shock staged by the OPEC countries had a substantial and greater effect on the mind of the Japanese businessman than his counterpart in the West. The Japanese tradition of looking ahead to the effect of decisions on the next generation, rather than next year's balance sheet, had meant that from that moment energy efficiency had taken a much greater role in Japan than elsewhere. As a result, the cash-saving 'no regrets' options available to the other members of the JUSCANZ group had already largely been used up in Japan. In other words, to reduce emissions Japan would have to incur real costs to its economy, not just the political inconvenience of the other industrialized countries. To a large extent Japan's attitude appeared to be that it was prepared to do more to cut its already relatively low emissions when the other industrialized nations caught up with the measures it had already taken 20 years before. It was therefore a member of JUSCANZ by default, in that it was not prepared to make new commitments for itself. It did not mean it would not have been happy to see its fellow industrialized nations do so, but unlike Australia the Japanese did not have the front to plead that they were a 'special case'.

The European position was also more complex than it appeared. As already explained, the Nordic countries and particularly those affected by sea level rise were very keen on reducing emissions. The Danes had gone in for big incentive schemes for windmills for alternative energy, and Holland were reducing traffic and increasing energy efficiency. In the south, Spain, Portugal and Greece were the 'developing' nations of Europe. They had demanded and been granted increasing carbon dioxide emissions in order to catch up with the development of other members of the European Union. Germany's position had been complicated by unification and the recession which had pushed the environment off centre stage.

Rather oddly, into this gap strode the United Kingdom. Largely thanks to domestic policies that had nothing to do with climate change, Britain was able to produce figures which showed it was well ahead of most other European countries in reducing emissions. The privatization of the electricity industry had brought about structural changes in the way power was produced. The closure of many of the coal mines and the coal-fired power stations which had dominated the energy industry meant that Britain would undershoot its target by as much as 12 per cent. This meant that many of the other measures like energy efficiency and 15 per cent VAT on fuel were not necessary. In fact, that one policy change and its extraordinary and completely unforeseen knock-on effects had completely changed Britain's predictions. Overnight, it transformed the United Kingdom from a reluctant participant in climate change talks to a leading light advocating action. Only politicians can take advantage of accidents of history like this and with a straight face claim credit.

In the case of Berlin it was John Gummer who, as environment secretary, took over as head of delegation. With the zeal of a new convert he made substantial contributions to forcing through a deal, and became a bit of a folk hero among the green groups, although not without some jokes at his expense. Some wondered whether, if there really was a God, he would at any moment be struck by lightning.

The secret of the European Union's continued ability to pledge to return emissions to an even keel by the year 2000 was based on offsetting the good performance of one country against the failure of another. It had always been intended that reductions in emissions in northern European countries were to make up for the increases in southern Europe. This would work, of course, only if all the northern targets were met, which for some countries seemed doubtful, but then Britain had unexpectedly turned up trumps. On the other hand there was a tendency to predict larger increases than actually occurred. Governments loved to forecast economic growth and therefore extra carbon dioxide emission increases when in reality it did not happen. However, for the purposes of Berlin, the European Union was able to hold firm on the basis that between all the countries involved it would be possible to stabilize emissions at 1990 levels by the year 2000.

The South's views on global warming were developing very rapidly at COP1. As has been stated, it had become clear that it was among the G77 countries that the first, and also the worst effects of climate change would fall. Not only would some of the small island states disappear, the less-developed countries also had fewer resources with which to adapt.

Sea defences were the simple example. In theory it is possible for the United States to build defences to save itself from the effects of sea level rise, and this is true of most of Europe. All the industrial countries could afford to adapt, at least until their economies were severely damaged in other ways. The G77 countries, on the other hand, faced disaster from the word go. By the time the delegates from Egypt, Bangladesh, Vietnam and China had reached Berlin they were asking themselves how could they defend their deltas from sea level rise? Some places could be protected fairly cheaply but vast food-growing areas were being threatened.

Between Rio and Berlin a number of key countries, including China and India, had decided that the global warming crisis was not just a problem of the industrialized countries but a world problem. For the developing world climate change was still a less urgent problem than poverty and the lack of means for development but it was moving up the agenda. However, they readily took up the rather doubtful claim of the industrialized world that first they had to develop economically to make enough money to deal with environmental problems. For many this is a false starting point since the two must go hand in hand, but G77 continued to make much of it. The fear persisted from Rio days that the environment might be used as a way of preventing the development they craved.

Indeed they were not far wrong. Early drafts of the European papers for the conference talked of the need to get emission targets pinned on the developing world too. Realizing that this idea would have caused a collapse of the talks, the passage was hastily removed. The United States, and particularly its industrial lobby, regarded this as a fundamental point, but bided their time and left it largely unspoken. At Berlin the possibility of getting the countries of the South to concede that at some time they must become part of the reduction process was premature. The difficulty was that the developing countries had neither the capital nor the inclination to do anything about climate change unless the countries who had caused the problem in the first place showed that they were serious about tackling their emissions. In these circumstances it was still vital to get the United States and others to acknowledge their responsibility, and then force on them meaningful targets. This was the reason for the G77's green paper and the switch in support to the AOSIS protocol. The developing countries believed that if the industrial world was serious about climate change it would accept the challenge of a 20 per cent reduction target by 2005. Remember that in scientific terms this remains a minimum first step. But in political terms at Berlin, and sadly since, it has remained an impossible dream.

One group, which nominally remained in G77 and had once been a large influence in its inaction on this issue, had now broken away and stood much more in the shadow of the United States. This group were the oil producers, particularly Saudi Arabia and Kuwait, so recently partners with the United States in the Gulf War against Iraq. Many times these countries had flexed their oil muscles because of their strategic importance and had successfully slowed down the climate talks in the run-up to Berlin. It is hard to be sympathetic with countries which have such fabulous wealth, sometimes so wasted and abused, but it would be wrong not to acknowledge the genuine concerns of oil producers and their role. For a start, the economic survival of some countries currently depends almost entirely on oil revenue. A carbon tax, once seen as the answer to cutting carbon dioxide emissions by making fuel more expensive, would clearly hit those exports hard. Some industrialized countries favour the idea of a carbon tax simply because it would be a useful way of raising large sums in indirect taxation. The United Kingdom's annual increase in the price of petrol, with global warming given as one of the reasons, is a sort of carbon tax. If, on the other hand, there were a different form of tax, a well-head tax levied by the producer countries, then oil exports and production might fall but revenues for the exporter countries would remain the same. This would have the added advantage for oil producers of conserving reserves. A possible future political solution if all parties are to be brought willingly to the negotiating table would be the introduction of both a well-head tax and a carbon tax, balanced to give both sides a bit of the action.

But those are still issues for the future. In Berlin these ideas were the stuff of economic theory, and what was needed at COP1 was a hard-nosed political deal. In reality new targets and timetables for achieving them were out of the question. The United States, with the right-wing Republicans on a bandwagon, just could not deliver. But that did not mean that a deal could not be done, or that the Clinton administration was not the key to making it, and secretly keen to do so. It is important to consider some of the competing forces which appear to lead to curious American behaviour at climate change talks and other environmental forums. This pattern has been observed time and time again. In this case there can be no doubt that senior United States scientists and their political masters are convinced of the global warming problem. In the days before the Berlin conference vice-president Al Gore had called resistance to climate policy 'intellectually, politically and morally bankrupt'. Yet here was the American delegation at Berlin besieged by their own industry group. The administration was also being carefully watched by the right wing in Washington, demanding resistance to a deal for all they were worth. However, at the eleventh hour on the last day something changes. When despair is setting in, and, crucially, Washington critics of the United States administration are literally asleep, a form of words acceptable to all sides emerges.

News that a deal appeared to have been done was a genuine surprise on that last day in Berlin. Tension was high in the conference hall. Those of us who had not spent the night in private meetings in the inner sanctum of conference, known as 'room 7', or in the corridor outside knew there was a deal. The news was passed rapidly from one group to another as thousands streamed in for that final session. We journalists were still trying to get hold of a text and grasp its implications as the delegations began to take their seats. There were still doubts. Would everyone agree a deal stitched up after a 20-hour marathon session in room 7? It was reported that Mrs Merkel, chairwoman of the conference, had announced that no one would be allowed to leave to go to bed until they had agreed a document, but not all the countries were represented.

Anyone, from those nations who thought the deal was too weak to those who wanted no deal at all, could, in theory, scupper the whole thing by refusing to accept a consensus. Remember there were still no rules of procedure, in effect every nation possessed a veto. The moment of high drama was at hand. There were supposed to be 118 countries present. Mrs Merkel, who had been up nearly all night, looked completely in control and equal to any emergency. She was to put the final document to the conference. All eyes would be on the oil producers and in particular Saudi Arabia and Kuwait. We all knew the deal could be undone by one single voice of dissent.

Delegates, many weary from spending all night arguing over single words in the text, had filed in slowly. Twenty minutes after the meeting should have started everyone seemed to be more or less in place. A few people were lounging about chatting. Mrs Merkel said: 'You have all seen the final document.' This comment was translated into half a dozen languages. 'Any objections?' she asked. Seconds passed. The conference seemed half asleep; did they really know this was the moment? Nobody stirred, and then down came the little hammer. 'It's agreed,' she said, breaking into a broad smile. It had been a tough few days. There was suddenly a stir in the Saudi Arabian camp. Had they realized too late? An urgent consultation went on but no one stood up. The conference passed on to procedural matters and self-congratulation. At last we had a story, history in the making. After all that talk of failure there was a deal. Whatever it was, it had gone through. Time to find out what it said, and more important what it meant.

The Berlin Mandate, as the agreement was called, had kept the climate show on the road. Remember this was the first meeting of the parties (COP1) and the Convention on Climate Change allowed for an annual meeting thereafter. The ratchet effect built into the treaty was working. In Berlin, no one believed that COP2, to be held in Geneva on 8–19 July 1996, would be such a media event. As it turned out the science and politics had moved on. The US Democrats had decided to make climate change an election issue and the ratchet took another turn, setting the stage for tough targets to be argued about at COP3 in Japan in December 1997.

Apart from the tricky wording of the final document we should also deal with the basic machinery set up at Berlin. An administrative centre was needed so that the Convention could be kept up and running between meetings. COP1 agreed a secretariat would be established in Bonn, a feather in the cap for the Germans who would have plenty of spare office space when their government moved back to the old capital of Berlin. A budget of £12 million was voted, enough money to fund a staff of 43 over two years until the Japan meeting, when more money would be needed. Two subsidiary bodies were set up, as provided in the treaty: one to help with scientific and technical advice and the second to supervise the implementation process.

So to the Mandate itself. Some of the language was tortuous in order to avoid politically sensitive words like 'targets' and 'timetables' but it meant the same thing. The Mandate agreed first of all that the measures agreed to tackle emissions were inadequate and that 'appropriate action' was required beyond the year 2000, including the strengthening of the commitments of the industrialized world.

These countries had to detail policies and measures they were to take and 'to set quantified limitation and reduction objectives within specified time frames such as 2005, 2010 and 2020 for their anthropogenic emissions by sources and removals by sinks of greenhouse gases.' This was the key paragraph because although it avoided the words 'targets' and 'timetables' it included dates and words which could hardly mean anything else. Paragraphs in the Mandate said that the process would be carried out in the light of the best available scientific information. Early analysis and assessment of policies and measures would be made for the industrialized world so that results could be achieved with regard to time horizons such as 2005, 2010 and 2020.

> The protocol proposal of the Alliance of Small Island States (AOSIS – proposing a 20 per cent reduction in industrialized country emissions by 2005) should be included for consideration in the process. The process should begin without delay and be conducted as a matter of urgency to ensure the completion of the work as early as possible in 1997 with a view to adopting the results at the third sessions of the Conference of the Parties.

So the agreement specifically called for new measures from the villains, that is those countries who had created the climate problem in the first place, and gave them a strict timetable to come up with a new agreement. It said a legally binding deal should be ready to be agreed by COP3 in Japan.

The agreement excluded any new commitments for the developing world beyond a general instruction to keep greenhouse emissions to a minimum which appeared in the original treaty. But it therefore continued, however, existing responsibilities, which meant that the developing countries had to provide during 1997 details of their existing emissions and the plans they had to combat them. This so-called baseline information, which would subsequently be checked by visiting expert teams, is seen as a first stage in drawing them into the treaty process. The 37 industrialized countries and countries with economies in transition (that is the former Communist bloc) had to submit these plans and have them reviewed by the summer of 1996. By the third quarter of 1995, 27 of the 37 had produced them, and seven had been reviewed, including the two biggest, the United States and Japan. Progress was being made, although the chances of both these processes being completed on time were remote. At the time reactions to the deal were mixed. My reporting of the event had 'the world signing up to tackling climate change after last minute concessions by the United States and OPEC countries led to a breakthrough.' It also said: 'The final document is a fudge but for the first time clears the way for negotiations on targets for cutting carbon dioxide emissions.' It recorded that the final deal had been sealed by Mrs Merkel at 6am after two all-night negotiating sessions.

But principally the *Guardian* prefers reporters to quote other people's reactions to the deal, rather than their own correspondent's views. These were remarkably varied. Reactions by politicians and lobbyists, now often called 'soundbites', are often made within minutes of a decision. Below are some of those gathered by me on that day.

Tuiloma Slade, speaking for Western Samoa, one of the countries threatened with extinction by sea level rise, said he was deeply disappointed that the document did not specify 20 per cent cuts in carbon dioxide emissions by 2005:

> We will campaign relentlessly for cuts as soon as possible. The strongest human instinct is not greed, it is survival. AOSIS countries will not barter the survival of their culture and their homes for the short-term economic gain of others.

India's minister T. P. Sreenivasan said:

To have failed here would have been a failure in the eyes of the world. But we have resisted attempts by some groups to amend our efforts out of all recognition. It means the developing world can still grow to meet our social and development needs while the developed world has acknowledged the need for deeper cuts in greenhouse gas emissions.

Pleased with his own role in the negotiations, John Gummer for the UK commented:

The deal means we have a real chance of avoiding the worst of climate change. It is the first and major step to reducing the emissions which are causing the problem. We can have some confidence that we will be able to protect our children's future.

The United States' reaction was more muted. All Timothy Wirth would say was that he was 'satisfied' with the outcome and that his government had accepted it must produce new measures to cut existing emissions. Considering the pressures he was under it was not surprising he kept a low profile. Here are two reactions from opposing American lobby groups.

The United States' oil and coal lobby was glum, and the Sierra Club, the oldest and most venerable of the American green groups, was angry. First the Global Climate Coalition, which represented fossil fuel and chemical interests in the lobbies in Berlin, said in a statement:

The agreement gives the developing countries like China, India and Mexico a free ride. It could hit American consumers in the wallet by raising the cost of energy supplies such as gasoline and electricity, while at the same time costing an average of 60,000 jobs a year due to loss of competitiveness in global markets.

The Sierra Club:

The weak Berlin Mandate only postpones action that world leaders should have taken today. The US administration stands shoulder to shoulder with OPEC as they take the brunt of the blame for this delay. The United States has the resources and technology to slash global warming pollution, yet the US performance here was a shocking abdication of leadership.

The Clinton administration's stalling tactics forced an inadequate resolution to come from this meeting. Rather than acting now, 150 governments will spend two years yammering about the treaty with no clear guidance as to where they will end up.

Tessa Robertson for Greenpeace International said:

This was a missed opportunity to get real reductions agreed now but it is a better deal than we could have expected two days ago when there was a real chance the whole deal would collapse completely.

Perhaps the man with the greatest insight on the day was Richard Benedick, an American, who was a veteran of the 1980s negotiations on the Montreal Protocol on ozone depletion:

I am delighted. It may not seem such now but a process has begun that is politically unstoppable. No one can pull out and each meeting will bring harder targets and tougher timetables.

For all the myriad environmental groups the daily newspaper *ECO*, which is a feature of all these conferences, summed up:

When ministers finish congratulating each other and the dust clears, the stark reality of the global predicament will remain unchanged. The build-up of carbon dioxide in the atmosphere will not have been slowed by the talks here. The concentration of greenhouse gases in the atmosphere can only be stabilized at non-dangerous levels if global greenhouse emissions are reduced well below 1990 levels. This will be accomplished by changes in energy consumption patterns, a shift to renewable sources, and a reversal of deforestation, not by talks in Rio, Berlin, or Kyoto [Japan – the venue for COP3]. Commitments, existing or new, are worthless if they are not implemented.

To breathe life into the climate treaty process, governments must get serious about implementing emission cutting policies – starting immediately.

It seems to me with hindsight that there is some truth in all these instant responses. In one sense, considering the size of the problem and the crisis facing the world, it was a vague political document with no teeth. However, given the political realities of a world still in recession with the dogma of market forces and short-term profits taking precedence over even the medium-term future, it was a deal that kept the process alive.

Since Berlin many meetings have taken place and, as already discussed, the science has been moving on apace. In this political section we have freely quoted some environmental and industry groups, obviously pulling in different directions. Their roles in past talks and the future of this Convention cannot be overstated. Both claim to represent the ordinary people, and that has validity. While the fossil fuel lobby has clearly been to the fore in the past, other industrial forces are awakening to the dangers that climate change poses. In the next chapter we examine some of these opposing forces. A controversial issue already referred to, which was particularly important in Berlin, is joint implementation or JI. For such an apparently dry subject it has great potential to ruffle feathers. There is a shortish chapter to explain this issue, which will surface again in negotiations. We can then get back to what we have achieved since Berlin, and examine where the world might go on climate change from 1997 onwards.

# 16

---

# Pressure groups
# pull both ways

POLITICIANS CONSTANTLY INSIST that they act on behalf of their constituents but a good number of voters never get to put their point of view to their representative except through the ballot box. Instead politicians are bombarded by information from pressure groups. Very often these spring up spontaneously in response to a single issue, for example to oppose a new airport or motorway, but more commonly pressure groups are long-standing organizations that exist to support various causes. Many of the most recent, largest and most successful pressure groups to arrive on the political scene campaign on environmental issues. In Britain most people will have heard of Greenpeace, Friends of the Earth and the World Wide Fund for Nature, which, between them, already have a larger membership than the three main political parties. These are multinational organizations in their own right, challenging governments and industry in many countries at once. Their message to their members is that humanity is wrecking its own environment. As a result, the fate of the elephant and the panda, the halting of pollution and overfishing, or in this case saving the climate is in our own hands, if only we were prepared to be more sensible.

This is a powerful message and essentially non-political, since no politician would admit to being 'anti' the environment. However, since it is only by political action that problems caused by human activity can be solved, politicians are constantly under attack for not doing enough. All three organizations, along with many others, are involved in climate change campaigning. It is after all *the* big issue even if it might not be the most glamorous or attract the most donations. There is far more money in slogans about saving whales and elephants than in climate. However, campaigners work on the issue of climate full time and also support umbrella organizations like Climate Action Network, which operates internationally. The message is roughly the same from all of the groups: we can and must slow climate change down. We must because in doing so we can save millions of lives and vast areas of the natural world which would otherwise disappear. They do not add the nightmare scenario which follows that logic. If we do not slow the process down and give ourselves time to adapt it is probable that civilization in large parts of the world will collapse. This lack of scare tactics is called 'responsible behaviour' by the pressure groups, who are wary of being returned to the loony fringe to which they were once consigned.

One of the arguments that used to be central to the debate but is hardly heard now is about growth. The so-called dark greens, readers of the *Ecologist* magazine, which is tough on capitalists, have always questioned the idea that economies have to be in a constant state of growth. Their argument is that constant growth is simply not sustainable and therefore at some point the people in the industrialized countries will have to live a simpler lifestyle, consume less and recycle more materials. The counterargument is that without growth there is no extra money to pay for environment improvements. No doubt the argument would have continued longer if the Third World countries had not sided with the capitalists. They made it clear from the beginning that economic growth was their number one priority in order to rid themselves of poverty. There was then an immediate hasty retreat by the greens on the growth question. Principally, they did not want to upset the developing world but were also wary of being labelled as crazy for wanting to lower standards of living by opposing continuous growth.

But however responsible environment groups want to be, their job is to bring unfortunate facts to people's attention, information which is generally scary and uncomfortable. This often leads to confrontation because many people, for example industrial polluters, would rather not know about it, or at least not have their actions broadcast. Among the groups who would rather not know uncomfortable information are politicians. After all, stopping activities that create jobs or make profits means taking some rather difficult decisions and risks losing votes.

But politicians are responsive to public opinion, and if pressure groups marshal it sufficiently then action may result. The problem is that in climate change there are two lots of lobbyists pulling in different directions, and the second group has a distinct advantage. Few of the general public have heard of the Global Climate Coalition, the Climate Council, or a man called Don Pearlman, but so far in the global warming debate these unknowns have been at least as successful as the green groups in bending politicians to their will. Time and time again they have effectively stopped both scientists and politicians taking action to combat climate change. They also have an in-built advantage which is necessary to explain.

Researching this book has entailed a lot of reading. Two of the books consulted on environmental politics touch on this point and both authors feel it necessary to quote the fifteenth-century Florentine statesman Niccolo Machiavelli, to try to explain the problem. He knew a few things about politics and human nature. Tony Brenton in his book *The Greening of Machiavelli* compares the modern states' behaviour on the environment with Machiavelli's advice to the Prince. He says the nation state amorally pursues its own interests, and expediently sets aside any concern for the global good. In fifteenth-century Italy the product of such politics was an era of warfare and betrayal among petty states. As a result they neglected the very real external dangers they jointly faced until they were consumed by them. The second author, Ian H. Rowlands in his book *The Politics of Global Atmospheric Change*, chose a direct quote. For my purposes it explains why the greens are struggling so hard to get their point across and the industry lobby has an in-built advantage. Over to Machiavelli:

It must be considered that there is nothing more difficult to carry out, nor more doubtful of success, nor more dangerous to handle, than to initiate a new order of things. For the reformer has enemies in all who profit by the old order, and only lukewarm defenders in all those who would profit by the new order, this lukewarmness arising partly from fear of their adversaries, who have the laws in their favour; and partly from the incredulity of mankind, who do not truly believe in anything new until they have had actual experience of it.

175

Despite the fact that few have heard of the pressure groups created by the oil and coal interests to fight their corner there can be no doubt that they represent the most powerful industries in the world: coal, oil, and automobiles. In this case they also have the message politicians most want to hear – do nothing at all; continue with business as usual; any action to combat climate change will damage established industries, and millions of workers. To back up their campaigns they have unlimited rersources which they have used to good effect since before Rio, realizing what a threat the Climate Change Convention was to their interests. Fearful that politicians might take decisions which will damage their profits by cutting consumption of fossil fuels they have been paying teams of lobbyists to work on their behalf. At every meeting anywhere in the world where climate change is to be discussed the oil industry is there. At first sight it does not look like the oil or coal industry. As we have already said, they call themselves names that could mean just about anything in order to become a non-governmental organization or pressure group, which entitles them to attend the talks, and set to work. The Global Climate Coalition, which among others represents the automobile industry and coal, uses the slogan 'Growth in a global environment'. Their brief is simply to slow down the business of doing something about climate change as much as possible. It is a war of attrition, or as Adam Markham, head of climate change campaigning for the World Wide Fund for Nature, puts it, 'These guys play hardball.' Even though the Coalition could claim moderate success at Berlin in slowing down action, their response was to complain that the developing countries were not being asked to do anything and that 'US jobs, economic activity and international competitiveness are at grave risk'.

The Coalition's headquarters are in Washington, also the base for an even tougher customer, Donald H. Pearlman. There was great delight in Berlin when the respected German magazine *Der Spiegel* ran an article exposing his activities. Within hours it had been translated into English and distributed to delegates. Headlined 'HIGH PRIEST OF THE CARBON CLUB' it detailed his operations, saying that: 'Successful as no other lobbyist in international policy, he directs his worldwide climate network – always to block.' It recorded his lobbying at Berlin, including 'his orders' to Kuwaiti delegate Atif Al-Juwaili as he entered the conference hall. One morning he was alleged to have told him that 'We urgently need someone on the Bureau'. A few hours later, the Kuwaiti delegates were energetically demanding a seat on the future UN bureau, which oversees the implementation of the Climate Convention. (They later succeeded.)

The article continued:

> Pearlman has worked hard for three years to ensure that climate protection negotiations end in the never-never-land of vague declarations. He has not missed a single one of the over twenty scientific and political conferences leading up to Berlin. Hardly anyone is as familiar as he is with the more than 1000 UN documents.

The magazine goes on to say that he is a partner in the Washington law firm of Patton, Boggs and Blow and keeps his clients secret. Among the 1500 companies the firm represents are the industrial giants Exxon, Texaco, Shell and DuPont.

> Pearlman sees himself as a preserver of the American way of life and identifies himself closely with the US establishment. America's wealth is based on the oil industry, so that is the way it should stay. The permanent task of his crew is to watch over US diplomats. 'Every word we state publicly here appears on the desk of members of Congress the next day. These people create an atmosphere of mistrust and suspicion,' complains a US climate expert about Pearlman and his helpers.
>
> To undermine the credibility of scientists, Pearlman systematically uses representatives of the Gulf States at meetings of the Intergovernmental Panel on Climate Change. In September 1994 in Geneva, during the decisive meeting (for the influential scientists' report to Berlin) Pearlman and his Arabic friends brought chaos upon the international gathering of experts. 'We only wrote down undisputed statements,' reports the Dutch climatologist Joseph Alcamo. But the Pearlman alliance, as Alcamo angrily recounts, 'questioned every single line in the report', and engaged in 'endless hair-splitting'. Magically, the lawyer turned himself into an expert. Representatives from Kuwait submitted proposals to change the original text – in Pearlman's handwriting. Not even the sentence that 'an increase of carbon dioxide is expected in the atmosphere' went through. Triumphantly, Pearlman announces

'there is no scientific consensus' on the threat to the climate, even the IPCC is debating the issue.

*Der Spiegel* goes on to accuse the 'coal fundamentalists' of engaging their own scientists and then coming up with dubious findings to support the case that there is no evidence that extreme weather events are occurring. 'With such disinformation the US oil lobby dictates the public debate,' says Christopher Flavin, vice-president of Washington's environmental group the Worldwatch Institute.

Merylyn McKenzie Hedger, who has consistently attended both the IPCC meetings and the political sessions on behalf of the World Wide Fund for Nature, with the aim of exerting pressure in the opposite direction, has seen Pearlman in action. Since the Berlin meeting Pearlman has kept up his 100 per cent record of attendance at every session of every committee. She said:

> He is there arguing over individual words in the text, watering down the science so there are more uncertainties. The scientific consensus on climate change is unprecedented in world history. Never have so many scientists agreed on so much. Yet listening to Don Pearlman and company it is hard not to believe there are disagreements over many things, and as a result of his efforts qualifications have been added to the text. Having succeeded in this ploy his lobby can and do argue in the next meeting and all the ones afterwards that the science is uncertain and therefore it is too early to take action. Instead, they say, more research is needed to iron out the uncertainties, so that action when it is taken can be in the right direction.

Another ploy is to make the document more obscure. Scientists are very keen to include graphics to show how carbon dioxide is increasing in the atmosphere or other simple illustrations. This is particularly valuable in summary documents released to the press and politicians so they can better understand the issue. Time and time again Pearlman and his associates object to graphics in reports and several times they have succeeded in having them dropped altogether. Along with other journalists I have tried to put these allegations to Mr Pearlman but he declines to talk.

Watering down official documents and slowing down negotiations are not the only tactics employed. Using some of the fossil fuel industries' vast profits to fund sympathetic scientific research can also be fruitful in the propaganda war. Funding appears to be hard to find unless you happen to be a scientist who casts doubt on the global warming theory.

If one legitimate scientist can be found to argue against the global warming theses of 1000 other scientists he or she will get research grants and funding – and when the work is done you can be sure the results will be heard by the politicians. This does not mean that the work is not genuine. There are many uncertainties in climate change as the IPCC points out. For example, using computer models it is possible to factor in extra cloud cover to take the sting out of the predicted temperature rise, or higher snowfall at the poles to discount the expected sea level rise. It is even possible, using some calculations, to show that there will be global cooling. This is not what the vast majority of climate scientists think will happen, but it is not possible to prove it is wrong.

These 'minority reports' receive plenty of attention. They get coverage in newspapers and on television, partly because editors like reporting perverse theories and it often suits the politics of the paper, but mainly because the resources and the expertise of the fossil fuel lobbyists have been used in serving up the research in the form of useful press packs and film material. It is an easy story to cover because nearly all the work has been done for the lazy journalist. It takes a strong mind to reject a story which sounds exciting but in truth is manufactured to suit a powerful lobby and the politics of your newspaper. By contrast, the consensus science, from the official sources, is much less accessible to journalists and often has to be ferreted out from piles of documents. The language of these papers is as the scientists wrote it, and sometimes made even more obscure, if that is possible, by the intervention of the fossil fuel lobby.

Despite this, read as a whole, the Intergovernmental Panel for Climate Change documents paint a stark picture for the future of humanity. But it is hard to find someone who has read it all. The reports run to thousands of pages. Even worse, some scientists are so careful not to appear as scaremongers that they write in a totally incomprehensible style. The prediction of the future which everyone ought to hear loud and clear is written in a language which speaks to fellow scientists, not to the world in general. It is also a good bet that the experts on glaciers, snow and ice only read that section, and perhaps the bit on mountains. They almost certainly will not worry about the section on forests or deserts. It is, however, the whole picture which is so frightening, each adding an increment to the alarm. Hence the need for environmental lobbyists to translate the message into everyday English to press the points home. Up to now, with a far weaker scientific case, the fossil fuel lobby has done much better than the environmentalists in getting their way in the political process. It is not that the world has failed to be alert to the threat of climate change. It has been reported widely, and we have all noticed droughts, mild winters, exceptionally warm months. People love talking about the weather but have they really understood what climate change means?

The fossil fuel lobby must be given credit for grasping the fact that what matters in the real world is not the facts but the message that gets across. It realized early on that people are attracted by the fantasy of a Britain in which hot sunshine is guaranteed every year and in which grape vines cover the hillsides. If that aspect could be emphasized, and the tropical diseases this extra temperature might bring downplayed, then global warming would not be seen as a problem. It is not, therefore, the fact of global warming that matters, it is the gloss that is put on it.

The environment movement took some time to realize that it had been outflanked. It had talked to the media, and in doing so felt it had done its best to inform the public. For the purposes of getting agreements pushed through at local and international level lobbyists had mostly relied on direct approaches to politicians. This rather unimaginative method was surprising because in other areas of campaigning there has been greater success in making their side known to a wider public – particularly in getting the press and television favourable to one point of view. As already said, this has an effect on public opinion whatever the rights and wrongs of an issue. In other words, what the voters think matters at least as much to politicians as any kind of logical argument at the door of the conference centre. It may sound cynical but the ability to manipulate the press has meant the difference between success and failure in many green campaigns, especially when it comes to real political reaction. Below are two examples where the green lobby has attempted with great success to outflank industry and government. In climate change other options to combat the oil lobby are now being explored successfully, and give great hope for the future, but we come to those shortly.

The simplest illustration of successful manipulation is from the campaigns of Greenpeace. Over 25 years the organization has synchronized its 'actions' on boats all over the world with newspaper and television coverage. The long-running campaign against the French over nuclear testing is an example and in Europe the prevention in 1995 of the dumping of the Shell oil platform, the *Brent Spar*, in the Atlantic. They both show talent for the manipulation of public opinion towards a legitimate political goal. The tried and tested method of Greenpeace is to explore the facts of a case and the science well in advance of going public on the campaign. Briefing papers are prepared and sent to all journalists. Reporters from influential papers and television crews are offered accommodation on Greenpeace ships and off they go, either to the French nuclear test site or, in the case of the *Brent Spar*, to see activists climbing aboard in the middle of the North Sea to occupy it. Even though Greenpeace is now a large multinational organization in its own right, the David and Goliath image persists. The pictures of young men and women climbing a vast oil platform to make their point or being arrested by axe-waving French commandos immediately enlists public sympathy.

After years of accepting these opportunities for filming and easy pictures the BBC has recently reviewed its policy because it felt it was being manipulated by Greenpeace. In a sense it must be true – there is no such thing as a free lunch – if you are living on a boat with people all of one point of view it is hard not to be affected by it. In addition it is hard not to admire those who volunteer to put their lives at risk for a cause. From personal experience I can vouch for the simple fact that it is harder to be horrid to them in print.

A second factor, and this was the BBC's main point, is that the film is likely to be one-sided. After all, while the law may be on their side, it is hard to sympathize with Shell using firehoses on people hanging on to the rail of a ship to save themselves falling into the Atlantic or a paratrooper putting an axe through equipment on the bridge of the Greenpeace ship. A journalist is supposed to be able to see through all of this and find the facts. The trouble is the sympathy of the public is attracted by the underdog, right or wrong. Shell and the French authorities were losing the propaganda battle from the word go. In both these two campaigns it can be argued that Greenpeace made some bad mistakes. On the *Brent Spar* campaign they got some of the science wrong about the quantity of toxic cargo. In the South Seas they lost both their main ships the *Rainbow Warrior* and *Greenpeace* to the French in one day, damaging their own ability to continue the campaign. But in both cases they had already influenced public opinion in their favour so much that neither Shell nor the French could recover all the lost ground.

But in climate change negotiations that kind of swashbuckling bravado is not available to Greenpeace or anyone else. During the Berlin conference Greenpeace climbers spent several days up a giant power station chimney, but outside the *ECO* conference newspaper produced by the environmental movement, and the odd press release jammed into journalists' hands, it hardly got noticed. In the end all the mainstream environment groups were stuck with patient lobbying in corridors.

Gradually green campaigners have come to realize that this is not enough. Radical new approaches and policy were required. Jeremy Leggett, science director of Greenpeace, a respected climate scientist in his own right, was the first to understand this and break the mould. He decided in the early 1990s that what was needed was to create other industry lobby groups to oppose the fossil fuel campaign.

Dr Leggett started talking to insurance companies, first of all to convince them that their own businesses would suffer because of climate change, and then to get them to do something about it. For him it has been a long haul, but his campaign in this area and its visionary approach to the future of climate negotiations have become a role model for other groups. What Dr Leggett has succeeded in doing is to create a lobby which is set to become as important as the coal and oil industry group but pulling the other way. So far they do not have a Mr Pearlman at the negotiations but they do have financial muscle they may use in different ways. Dr Leggett's original vision was to get two of the world's biggest industries locked in battle for the future of the planet, but as things have turned out he is doing better than that. While most insurers still slumber on, some of the brighter companies, now significantly joined by bankers, are beginning to realize that their businesses are in mortal danger and are planning to take action to save themselves. Before we go into details on that we must deal with one other aspect which has held the environment groups back on climate change – an excess of responsibility.

A charge of being too responsible may sound like an odd criticism, but environment groups have been badly stung by claims that they have cried wolf about the environment. It is an accusation frequently made by those with a vested interest, who want nothing done to inconvenience the status quo. Politicians, particularly conservative ones, have frequently accused environmentalists of scaremongering. Like all smears it has some basis in truth. In the early days when the facts about climate change were beginning to emerge, exaggerated claims by environmental groups damaged what was a legitimate case. For example, sea level rise is clearly a threat to coastal regions and in Europe there are many low-lying areas potentially at risk. Many of them are also heavily protected by sea defences. For example much of the fens of eastern England are at or below the high-tide mark through centuries of land drainage. The peat soil on which the fens rest has shrunk as the water has disappeared. In some places the ground is 2–5 m (6–16 ft) below recorded levels of a couple of centuries ago. If a simple view is taken of sea level rise of say a metre (3 ft), plus the effect of a storm surge in the North Sea, then it is possible to say that the whole of the fen country is in danger of flooding.

In that strictly simplistic sense it would be. However, this scenario fails to take into account the massive area of land involved. The whole area is covered with banks, holding in rivers and drains. There are literally dozens of old banks from sea and flood defences dating from Roman times. It is not possible to envisage a storm big enough to push up the massive volumes of water required to flood this area in one tide. Nevertheless, a storm surge could, in theory, push the sea level to 5 m (16 ft) above normal. So, if you were an environment group crying wolf, you could draw a map of eastern England around the 5 m (16 ft) gradient and paint it blue. This would put about one third of England under water and make Peterborough, 48 km (30 miles) inland, a coastal town – Peterborough-on-Sea. One such 'shock' report was produced by the environment group Ark in the 1980s. It was based on fact, but not on reality. At the time the environment movement in the United Kingdom was struggling for respectability. Their detractors were keen to paint them as alarmist loonies. The Ark report has lived a long time in the environmentalists' minds – long after it has been forgotten by the public. Many environmentalists were scientists, unable to get a traditional job, and many of them longed for the respect accorded to their mainstream colleagues. Peterborough-on-Sea was the sort of publicity that made most members of the environment movement cringe.

The problem for environmentalists is that even without any extra gloss the threat of climate change and the potential effects of sea level rise are both dramatic and potentially very expensive for society. Yet they still shrink from pushing the alarm bells too often and too hard. But if the threat of global warming is as real as the scientists say, then what the environmental groups should be asking for is pretty drastic action. But let us go back to the beginning of this chapter; this is what places them at a disadvantage compared with the coal and oil lobby. The environmentalists want positive and difficult measures which will affect lifestyles – the fossil fuel lobby want nothing to happen at all. If business is continued as usual they will see more cars burning more oil, more fossil-fuel-fired power stations, more business and more profits.

For those in power, the prospect of doing nothing about a problem that may not manifest itself for another few years, or at any rate not until after the next election, has tremendous attractions. No action means no offence to the voters, who are not reckoned to like change anyway unless it is for their own economic advantage, which most action on climate change is unlikely to be.

Politicians are anxious not to impose taxes or penalties on voters to pay for something as complex and as long term as global warming. They believe, probably rightly, that most voters barely understand the issue anyway, at least not as well as an extra tax on petrol or a rise in the price of electricity. While we are being rude to the general run of voters let us rub in the point about most politicians. A modern Machiavelli might say what appears to matter most to politicians is not what is right or wrong for the country but getting elected, or re-elected. After they have gained power they are sometimes prepared to make difficult decisions – but then only if they think the effect of them might be forgotten again by the next election.

So, as Dr Leggett concluded, this is pretty thin soil for the sowing of long-term measures to help the climate. What he thought was needed was to create a different kind of pressure, and he set about doing it.

# 17

---

# Insurance spawns
# new tactics

ALL REALLY GOOD IDEAS are simple ones, and Jeremy Leggett must be given credit for the best brainwave the green movement has had on how to combat global warming. He started his campaign in a straightforward way. As the director of science for Greenpeace in London he gave lectures to sceptical insurance company executives at seminars, where his appearance was no doubt regarded as a bit of light relief in an otherwise heavy programme. Quite soon, as he began to educate insurers about the risks they faced, he started to be taken seriously. The spectacular storm of 1987, which flattened a million trees in southern England and damaged many homes and offices, cost the insurers dear. Less newsworthy but equally painful for these same companies were the heatwaves and droughts of the 1980s, now being repeated in the 1990s. The simple fact that London is a city built on clay was bringing in thousands of claims – in Dr Leggett's view it was merely the tip of the global warming iceberg.

In essence, insurance is a simple business. Companies charge annual premiums based on known risks. As far as the weather is concerned this means averaging out the claims made over decades for various extreme events and setting the premiums accordingly. However, if something changes the rules and the weather suddenly gets worse, the claims multiply, and too late the premiums are seen to be hopelessly inadequate. If the calculations are too far out bankruptcy looms.

Take the case of London and clay. As anyone who lives in London

185

or anywhere else with clay soil knows, after a few weeks without rain cracks appear in the soil. The longer it stays dry, the bigger the cracks. Normally, in a climate like that in England it rains often enough to prevent that being a problem, but not any more. Between 1975 and 1993 the national bill for clay subsidence totalled £2.5 billion. Problems began mainly in the long dry summer of 1976 but were repeated again in several years in the 1980s and a new crop of claims has emerged as a result of the very hot and dry 1995 summer. There is now a Subsidence Advisory Bureau, based in Bexhill-on-Sea in East Sussex, which will advise on what to do. It says that if you have the bad luck to be built on clay it is best to keep trees and shrubs away from the house. Broadleaf trees can drain up to 50,000 litres (11,000 gallons) of water a year from the surrounding soil. Willows are apparently the worst offenders and home-owners are recommended to keep them at least 40 m (130 ft) away from houses. It advises that if cracks do appear in your house you should wait for rain and hope that the re-swelling of the clay will cure the problem. If you (and the insurance company) are lucky then the cracks will disappear again. If not then the house might need extensive and expensive underpinning. The idea is to provide a stable concrete raft for the house to stand on, or at least the damaged part of it. Otherwise, come the next drought or two the process will be repeated and the clay will eventually pull the house to pieces.

For insurance companies one of the straws in the wind is that many of the houses which were damaged by the clay in the droughts were Victorian. In other words they had stood quite happily on clay foundations for as much as 100 years without damage and then thousands required major repairs in a few short years. In response to this many companies, including General Accident, have now introduced a differential premium for building insurance. If your home, office or factory is built on clay soil insurance will cost you more. This could be described as England's first global warming insurance premium.

But putting up premiums was not what Dr Leggett and Greenpeace were about; they wanted the insurance companies to exert political pressure. In 1994 Greenpeace produced a report called *The Climate Time Bomb*. It detailed the huge number of natural disasters which had occurred since the beginning of the decade. It gave a sort of league table based on the size of the cost to the insurance industry. It is worth noting that natural disasters in developing countries rate only a couple of mentions, although the loss of life there was far higher. What matters to insurance companies is the bottom line, how much they have to pay out.

For 1992 the top of the league table read:

Hurricane Andrew, US: £16.7 billion
Cyclone Iniki, Hawaii: £1.7 billion
Typhoon Polly, China: £1 billion
Typhoon and flood, Japan: £0.8 billion.

In 1993:

Flooding United States Midwest: £8 billion
Gales US, Canada, Cuba, Mexico: £3.3 billion
Flooding Germany, France, Netherlands, Belgium:
   total £1.3 billion
Flooding Alps: £1 billion
Typhoon Yancy, Japan: £0.7 billion.

For the first time it was clear that the message was beginning to get across. The report was endorsed by Dr Peter Myer of the Swiss Reinsurance Company, the first group to take action on global warming seriously. These reinsurance companies tend to pick up the tab from big disasters and some had already gone out of business because of the sequence of storms. It was one of the factors that contributed to the huge losses of Lloyds of London. When the report was published the executive vice-president of Tokyo Marine and Fire, Shiro Horichi, said:

> The fact is that in recent years natural disasters, whose return period used to be regarded as at least 100 years, have transpired every year in various places in the world. It seems difficult to believe that these incidents are merely accidental.

The Association of British Insurers remained sceptical. A spokesman, Tony Baker, said: 'We are aware of the concern, but our research over 200 to 300 years shows there is nothing significant going on over that time frame. In the interests of policy holders we are still studying the emerging trends.'

Greenpeace, meanwhile, was busy sending copies of the report to all the countries involved in the climate change talks due in Berlin the following year, and to the major insurance companies. Dr Leggett used the launch to hammer home his point that global warming was a threat to the world economy: 'Banks, the insurance industry, and pension funds are all threatened. Between them they have one half of the equity in US business. If they begin to take this report seriously then we will see things happen.'

When asked about the many alternative scientific reports claiming global warming was bunk, Dr Leggett said:

> It reminds me of the early days of the anti-smoking campaign. For every scientist saying smoking was harmful there was one paid by the tobacco industry saying there was no evidence. The fossil fuel lobby is doing the the same today with global warming. Let us not forget that governments accept global warming is a danger and a fact. What we are saying is 'Let's have some action.'

By the time of the 1995 Berlin Climate Conference, COP1, at least some of the insurance world was beginning to see the problem Dr Leggett's way. In order to cope with big disasters the companies were hoping to introduce a claims reserve. The idea was that companies would put away a percentage of premiums into a reserve fund in order to cover unforeseen losses, at least a partial acknowledgement that Greenpeace might be right. It meant that senior executives had appreciated the dangers climate change posed in creating a spate of unexpected claims. Perhaps more important, the brightest thinkers in the insurance business had been assigned to looking at the possibilities of climate change and its effects on the business. One of these was Andrew Dlugolecki, chief manager of operations at General Accident, based in Perth, Scotland. He had been invited by the Intergovernmental Panel on Climate Change to be lead author on the effect of climate change on financial institutions. He was already writing the report recommending, among other things, that building design should be changed to withstand higher winds and other storm conditions.

He was also recommending a change in the land development rules so that new settlements, or individual houses, did not get built on land that would soon be subject to flooding, either through sea level rise or because of the increased rainfall predicted for Europe. In Berlin a seminar for insurers was held on the Sunday in the middle of the conference, chaired by Rolf Gerling of the Gerling Group, the German reinsurer. Dr Dlugolecki told his audience, 'Looking forward, I am sure that climate change will speed up, and I am sure that will have major implications for the industry.'

Ironically, while Mr Pearlman, on behalf of the fossil fuel lobby, was not far away attempting to convince the decision-makers that global warming was a myth, Frank Nutter, president of the Reinsurance Association of America, was telling his fellow insurers the opposite. He said at the seminar: 'Of the 25 largest insured catastrophies in the US 21 have occurred in the last decade, and 16 of those 25 involve a combination of wind and water.' If global warming meant an increase in natural disasters it could bankrupt insurance companies, he said. He was asked by one of the American religious groups which attended the conference, and had £60 million of investments under management, whether he had heard that insurers were switching their investments away from industries responsible for global warming. Mr Nutter said he had not, 'at least not yet, but I hope that happens'.

For the first time bankers had also appeared to find out about global warming. Sven Hansen, vice-president of the Union Bank of Switzerland, took the opposite line at the seminar from his government at the adjacent negotiations. He said that environment problems, of which global warming was the greatest, posed a threat to world financial stability.

Although the insurance world, including a team from Lloyds of London, kept observers at the climate change talks, they made few efforts in the lobbies. However, some observers did make themselves available for newspaper interview. Peter Blackman, assistant director of the British Bankers' Association, told the *Sunday Telegraph*:

> International banks are becoming convinced that the world faces serious changes to its climate, with equally serious economic consequences. The banks are increasingly becoming concerned at the economic consequences of climate change. It seems to me that there is a good case that we face significant and, perhaps, permanent changes in the UK and the world. It is also clear that climate change will have a dramatic impact on work and lifestyles of our business and personal customers; it could bankrupt some of them and make some of them homeless and jobless.

Despite the presence of these observers the 'carbon club', as it was known, still operated in the corridors, speaking for the oil, coal, auto and chemical industries. According to its leaflets, however, it represented the entire business world.

The International Chamber of Commerce, which was also supposed to speak for the insurance and banking industries as well as the energy industries, was also represented in Berlin but that too advocated no action on climate change. Only the green groups were active in asking for cuts in emissions.

Undeterred by the slow responses of the insurance and banking industries to the political challenge Dr Leggett kept plugging away. Following Berlin he wrote articles for magazines serving the insurance world hammering out his message.

> The worst case for insurers is very bad indeed. I have now personally been told by people at the top of the industry, in Munich, Zurich, London, New York and Tokyo, that climate change could bankrupt the industry, if the dice roll badly in the greenhouse gamble.

Dr Leggett said he thought that as few as two events could spark a worldwide insurance crisis – a category five (the worst kind of) hurricane in New York, and a drought-related wildfire taking hold in an urban centre. But such extreme events might not be needed; a string of smaller ones, or what one of the world's largest reinsurance companies, Swiss Re, describes as a 'machine gun fire' of catastrophes could have the same effect.

Already in the small island states the worries of insurers about the twin threats of greater storms and sea level rise meant that property insurance was no longer available or rates had risen to prohibitive levels. In a wider world lack of insurance meant building programmes would not be commissioned and businesses could not start up; others would be forced to close as insurance expired. The first casualties would be the insurance companies themselves.

Dr Leggett had already begun to link the bankers and the pension funds into the same doom-laden scenario. The threats to their investments were just as great. Imagine the effect on the value of returns on capital of the failure of the Thames Barrier to protect London from a storm surge in the North Sea, the impact of a super typhoon in Tokyo, or a wildfire in California. All of these are increased possibilities with climate change.

A week after Berlin, Dr Leggett was in Texas at the Solar Energy Industries Association annual trade fair, listening to an oilman give a presentation. It clearly cheered him up. Bob Kelly, executive vice-president of Enron, told the US solar industry that his company believed it could build 100 megawatt solar photovoltaic power plants to generate electricity at 5.5 cents per kilowatt hour – a price competitive with fossil fuel. Mr Kelly said he believed that global warming would drive investments in huge solar markets in the future. He wanted to place his company – the largest gas company in the world – firmly in the front line of the solar revolution:

> We think the multi-megawatt solar industry is ready to take off. Sometime our children will have to pay the carbon pollution cost. That could be a big number. It could be bigger than the budget deficit. We think there is a great big market out there, and we are going for it.

Another forward-looking industrialist at the conference was Philip Ludvigsen, vice-president of American Re, who was motivated by what he saw as a spectrum of environmental threats to profitability. The company has set up a subsidiary to help development of new environmentally beneficial technologies, what it calls forging business alliances for a better environment. Advice is free on risk management, access is offered to high-level contacts, marketing and sales. Once the project is deemed to be on the road to financial success then the subsidiary, AmRe Services, would invest. So far it had not found a suitable solar business but was looking for one since cutting carbon dioxide emissions would reduce the company's overall insurance risk.

Greenpeace has switched part of its campaigning strategy from pointing out what is wrong with the world's environment to finding solutions to solving the difficulties. Accordingly it has invested a great deal of time and effort in projecting solar energy as the way forward for the future. Dr Leggett, who had become director of Greenpeace International's Solar Initiative, was at the Willard Hotel, Washington DC in July 1995 for the joint Greenpeace and US Solar Energy Industry Association seminar. It was the first time that the industry and the financial institutions had come together to discuss global warming. All the major banking and insurance institutions sent representatives. Dr Jerry Mahlman, speaking for the National Oceanic and Atmospheric Administration, one of the top American climate research organizations, said that the right wing in the US Congress was hampering progress on the issue. He said: 'This issue has a relatively low level of scientific controversy and a vicious level of social controversy.' Of particular interest to the insurance industry was his comment that although there was little information on the changes in tropical storm frequency, there was every expectation that hurricanes, once formed, would reach higher magnitude.

Dr Dennis Tirpak, head of the climate change division of the US Environment Protection Agency, painted a bleak picture of the policy opportunities for the government in abating the greenhouse threat in the current domestic political climate. Carlos Joly, senior vice-president at UNI Storebrand, Norway's largest insurer, had some challenging points to make about insurance companies' investment policy and sense of responsibility:

> The fiduciary responsibility we have to our life insurance custom-
> ers extends beyond the highest investment returns. What good is
> it if we have a good return 20 years from now if we have ruined
> the world along the way as a result of our investments?

In time his company would be looking to switch investment from fossil fuels to renewables, though the relative scale of the industries precluded this in the short term, he said.

But despite these hopeful signs, Dr Dlugolecki, now an international expert on climate change, said early in 1996 in an interview with the author:

> What I have found in the insurance and particularly the financial
> services sector is abysmal lack of awareness of climate change.
> There is what I think is an alarming level of scepticism in general
> insurance. In the US insurers are not at all convinced that global
> warming is happening and will have an effect on them. Quite a
> few people in the UK think the same.

He says that scientists are beginning to advise that global rise in temperature will lead to more erratic rainfall. This is bad news for insurers because it will mean flash floods and subsidence from droughts, exactly as happened in 1995. Dr Dlugolecki is a member of the Department of Environment's UK Climate Change Impact Review Group, which is suggesting that it will also become windier in the British Isles.

> Property damage is really just beginning as far as we are concerned. You have also got to look at what the impacts on investment decisions will be . . . when the government starts to take action on energy consumption, what effect is that going to have on shares in energy companies?

Dr Dlugolecki suggested that investors should be encouraging fossil fuel industries to prepare for more climate-friendly activities. His work has already led to changes of policy at General Accident.

> It made us very determined about the need for differential rating for subsidence. It was in the late 1980s absolutely conventional for building rates to be uniform across the country. The advice I was getting from scientists was that drought was going to be more and more probable in the future.

General Accident insisted that building societies changed their computer rating to reflect the risk of subsidence across the country. Dr Dlugolecki is cautious about revealing other measures that General Accident has in mind to combat climate change but differential rates for flood insurance for vulnerable areas seems likely soon.

Despite his own belief that insurers need to be more active in defending their industry from the threat of climate change Dr Dlugolecki thinks it is unlikely that they will become aggressive lobbyists like the fossil fuel companies as Dr Leggett originally hoped.

> Most of them are still sitting on the fence. Even when they do realize the dangers they are in I think it is much more likely that their pressure will be subtle, but in its way far more effective. If companies are doing their job properly they will not be investing in business and industries with uncertain futures, and those are the fossil fuel industries, at least those who do not diversify and invest in new technologies. If insurance companies invest in solar energy, wave power and the like then their money will do the talking for them.

Towards the end of 1995 the British Bankers' Association circulated a report on environmental issues. It said in part:

> There are already parts of the world which are uninsurable due to climate change phenomena such as increased hurricanes, rising temperatures and rising sea levels amongst others. However, it is recognized [by the IPCC] that there are also implications for other parts of the financial sector, mainly banking and investment. Initially we were extremely sceptical about the potential impact of climate change on the banking sector. However, we quickly learned that there were real threats.

The report, by Berlin veteran Peter Blackman, argued that bankers might be successfully sued in the courts for assisting industries which were changing the climate.

> It will be difficult for parties to avoid liability for aiding and abetting processes and operations which breach environmental standards for the present and future which have been clearly stated, possibly been enshrined in legislation and/or regulation.

But the main effect would be the direct adverse physical one on banks' customers.

> Their operations will fail or become not viable. For instance, hurricanes in certain parts of the world are already destroying property, plant and machinery; temperature changes either way have a dramatic impact on industrial processes; reduced air quality impacts upon both industrial processes and population factors; and rising sea levels lead to coastal erosion and flooding. Some Pacific islands will disappear underwater before 2010 unless something is done to save them.
>
> More than half of banks' business customers are affected by environmental factors. Thus, more than half of bank lending is involved. Also a lot of bank lending today is for long terms up to 40/50 years and often for 20/30 years. Within the lifetime of loans granted today climate change is forecast to have a dramatic impact upon industrial operations within 20 to 40 years. Thus, enlightened risk managers are already assessing these sorts of environmental factors and there are cases in the UK and Europe where propositions have been declined because of these concerns

194

where the outcome is too uncertain to incorporate within a risk management analysis which needs clearly to identify, quantify and price all the risks.

In summary, the bankers thought that it was good to be involved in climate change talks because it would 'protect as far as possible existing businesses and safeguard opportunities for future business. Clearly there are enormous opportunities to finance new developments and the development of alternative energies, amongst others.'

After that rather startling summary, the report concluded rather limply that there did not seem much to be done at the moment, apart from keeping an eye on the problem and doing another report at the end of 1996. Compare that with disclosures in December's edition of *Harper's* magazine in the United States, which said:

The people who run the world's oil and coal companies know that the march of science, and of political action, may be slowed by disinformation.

In the last year and a half, one of the leading oil industry public relations outlets, the Global Climate Coalition, has spent more than a million dollars to downplay the threat of climate change. It expects to spend another $850,000 on the issue next year. Similarly, the National Coal Association spent more than $700,000 on the global climate issue in 1992 and 1993. In 1993 alone, the American Petroleum Institute, just one of fifty-four industry members of the Global Climate Coalition, paid $1.8 million to the public relations firm Burson-Masteller partly in an effort to defeat a proposed tax on fossil fuels. For perspective, this is only slightly less than the combined yearly expenditures on global warming of the five major environmental groups that focus on climate issues – about $2.1 million according to officials at the Environmental Defense Fund, the Natural Resources Defense Council, the Sierra Club, the Union of Concerned Scientists and the World Wildlife Fund.

So despite Dr Leggett's successes in bringing the issue of climate change to the attention of the insurance and banking worlds he still has a long way to go. Neither of these sectors possess the aggressive business instincts of the fossil fuel lobby. For the time being the green groups continue in a lobbying battle at each negotiating session, unaided by any industrial lobby. Not that this work is ineffectual: some green groups have long been included on the delegations of countries in the negotiations. It is partly because they have been pushing the politicians so hard that agreements have happened at all. Indeed some argue that the Framework Convention on Climate Change itself, signed by so many heads of state at the Earth Summit, only exists at all because of their efforts.

However, green organizations feel that the process of taking real action about climate change is progressing at a snail's pace. Indeed, the science of global warming has progressed much faster. We know now that it is a much greater threat that we realized five years ago, yet the politicians have not caught up. It was noticeable at Berlin that the Climate Change Coalition did not miss any tricks. An example was a leaflet from the United Mine Workers of America. It was a simple 500-word message on a handbill from 'An American Labor Perspective', containing an attack on the AOSIS protocol and the target of a 20 per cent reduction of carbon dioxide emissions on 1990 levels by the year 2005. The reasoning was much the same as that put forward by Donald Pearlman. They were asking politicians to do nothing, picking what they hoped was the most attractive option. But what carried weight was that it was signed on behalf of 200,000 active and retired coal miners in the United States. In other words by 200,000 voters, probably concentrated into a few critical constituencies. These are the sort of groups to which politicians are highly tuned.

This is what Adam Markham, a British citizen seconded to the United States World Wide Fund for Nature campaign on climate change, calls 'playing hardball'. He is a great admirer of Dr Leggett's campaign to involve insurers and bankers in the political process. Learning from that and the activities of the Climate Change Coalition he told the author: 'Industry groups are one thing, ordinary voters who could chuck you out of office are altogether another.' He added that it is here that the green groups have failed to act. He agrees that many voters have their own self-interest at heart when they enter the booth to vote, but it is not just economic matters that swing elections, environmental issues are creeping up the agenda. He says the trick is to turn that concern into enough political pressure to make a difference to a politician.

Mr Markham came up with two examples. In the United States

there is an organization called Trout Unlimited. It has a membership of 600,000. These are people whose idea of heaven is heading for the hills and spending hours in their waders trying to catch wild trout and salmon. It is not enough to have introduced species. There are people who spend all week working in cities to earn enough money to travel hundreds of kilometres to spend a weekend pitting their wits against the native brook trout. This is big business in several states; Idaho, Pennsylvania, Virginia, West Virginia, Maryland and Ohio are all the haunt of the brook trout. But the trout is a sensitive creature which cannot cope with pollution or temperature changes. The latest science on global warming suggests that the rivers in all six states will become too warm for the brook trout. If the species is to survive it will have to migrate northwards to rivers in cooler climes.

To Mr Markham this represents a political opportunity. For the 600,000 members of Trout Unlimited the loss of the brook trout habitat is a tragedy; for the hoteliers, fishing shops and tourist industry in those states the loss of the trout will spell economic disaster. This is clearly a lobby at least as potentially important as the coal miners. These are people that believe in the American dream, the open-air life, the right to hunt and fish. All this is going to be taken away by global warming.

On the same lines there are already problems along the Eastern Seaboard, where shooting Canada geese used to be a popular pastime in the autumn. For some as yet unknown reason the numbers of these migratory birds arriving on the coast have consistently declined in recent years until eventually a total ban had to be imposed on shooting them. No one knows why the numbers have slumped. Could it be climate change that is either changing their habits, or preventing their breeding success? So far no one knows but out there is a political lobby to be tapped.

Another lobby that has a powerful voice because it is concentrated in tight communities with concentrated voting power is the skiing industry. The one thing it needs is snow, but in the global warming scenario it is the seasonal snow in the mountains of Europe and North America which is likely to be affected by climate change first. Already further development of the ski industry in the Cairngorms in Scotland has been checked by the simple fact that it does not snow enough. It used to snow more often and keep colder for longer. This meant a reasonable season even on the lower slopes. There is now less certainty about the snow and it thaws more rapidly.

In parts of Europe, particularly the Alps, skiing and the associated holiday trade is a major industry. Artificial snow machines
are already in regular use to supplement what nature provides. If
predictions are right then a whole industry is in deep trouble.

The World Wide Fund for Nature consider the skiing industry
to be another potential political ally. But first it has to be educated
about its uncertain future. Then the anxiety that is generated has
to be turned into political pressure for the politicians to act.

Mr Markham said:

As I see it we have politically failed in the last two or three years.
Let's have no more Mr Nice Guy. Let's get to the front line people
who are going to be affected, people who are going to lose the
things they value. It may be the hunting, shooting and fishing.
It may be the skiing. We have to build the political pressure by
informing these people that they are going to lose what they value.
We have to get to them economically too by educating the banking
sector that their businesses might be a bad investment. If we tell
them that long-term mortgages are a bad idea because of climate
change, we are beginning to play hardball like the fossil fuel lobby.
If we don't get this campaign right now and start moving these
politicians there will be no turning back in 30 years' time.

# 18

---

# Red herrings
# and contrarians

IT WOULD BE BETTER for the start of this chapter if 'JI' stood for jolly interesting rather than Joint Implementation, but the fact is it does not. JI, as it is known in the jargon, did, however, cause some wonderful rows at the Berlin Summit and continues to simmer as an issue in the background. As already explained, joint implementation is a system in which one country, with advanced technology, does work in a less-developed country to reduce greenhouse gas emissions and then claims the credit itself, or in more sophisticated cases jointly with the recipient. Much was said in Berlin about JI by politicians, the industrialized countries being much in favour of it. The developing countries, on the other hand, were mostly against, seeing JI as just another excuse for the developed world to avoid their responsibilities.

It is tempting to skip over this subject, but if we persist, it is one that takes us to much more interesting areas. JI starts a trail which leads to people who are trying to avoid the painful reality that combating global warming means difficult policy changes in Western democracies. Potentially unpopular political decisions have to be taken to mitigate its effects. It also leads to the collection of pundits and experts called 'contrarians'. These people believe in all manner of technical fixes. They are prepared to check out any theory and support any argument which avoids action on climate change. In fact, the row about JI takes us to the centre of the political crisis which currently surrounds action on global warming.

JI has a history almost as long as global warming politics and gets a brief mention in the Climate Change Convention. The idea grows out of the perfectly sensible notion that technology transfer is a way of preventing developing countries making the same polluting mistakes as the industrialized world once did. In other words, for them it means a much cleaner industrial revolution. Equally well it can apply to schemes in one country that soak up the carbon dioxide produced in another. As early as 1988 Applied Energy Services, an American power company, decided to plant 52 million trees in Guatemala. The idea was that the trees would take up the same amount of carbon dioxide as that emitted by the company's new coal-fired plant opened in Connecticut. The Dutch state-owned electricity board has a similar scheme which charges consumers a small levy for overseas tree planting. If everyone adopted the same policy most of the world would be covered in trees, not a bad thing, but probably not enough to solve the problem.

The real argument here is about whether development in one country can be used by another as an excuse for doing less about its own emissions. It is worth recording the arguments because it is certain that in some form JI will play a part in the future of dealing with climate change. This is what Chancellor Helmut Kohl said about it in Berlin:

> When all is said and done, it is immaterial, for the climate of our earth, which country or factory emitted the carbon dioxide or other climate-damaging gas. In our joint battle against these greenhouse gases, we should therefore consider how we can achieve a substantial transfer of knowledge and technology to the developing countries and at the same time use the funds globally available for climate protection as effectively as possible.
>
> One promising way is 'joint implementation of measures' envisaged in the Convention. In many countries industrial facilities and power stations can be made considerably more efficient through modernization. This makes not only economic sense, but can also drastically reduce the emission of climate-damaging greenhouse gases. We can, of course, further reduce the remaining pollutant emissions of modern power stations too but such improvements are minimal compared to what we can achieve for out-of-date power stations using the same financial resources. For this reason, within the scope of joint implementation, we should give

incentives to the industrialized states, who will have to shoulder the greater financial burden, to undertake climate-protection investment outside their own countries. I think if these countries could set off part of these efforts against their obligations to reduce greenhouse gases, this would be in the interests of all states in the quest for effective global climate protection.

The Chancellor knew he was sticking his neck out. He immediately conceded this 'must not mean that the industrialized countries can neglect their own efforts at climate protection'. In order to sweeten the pill still further he suggested, 'Experience can be gained and mutual trust created in quite a short time by means of voluntary pilot projects.'

Japan, which had a good record on energy efficiency, probably had the best legitimate reason for endorsing JI. Sohei Miyashita, head of the Japanese delegation at Berlin, said that he thought it was crucial to start a pilot phase as soon as possible and that Japan was looking to help its neighbours with schemes through the Economic and Social Commission for Asia and the Pacific.

However, on behalf of the G77 group of developing countries, the Philippines totally rejected JI as a way forward. There were no obligations under the Climate Change Convention for developing countries to reduce emissions. There could, therefore, be no grounds for developed country parties to obtain credits by working to reduce emissions in developing countries. Instead they should start work at home where they had created the problem in the first place.

That was not the end of the argument, however. The fossil fuel lobby, and particularly the industrial groups from the United States, remained very keen on JI. They saw it as a way of building large power projects in developing countries, and the nuclear industry hoped it might prove a lifeline for an industry that had long been on the defensive. Not all developing countries were against it, either. The Declaration of Santiago, made on 8 March 1995, was an endorsement of JI signed by seven Central American presidents from Costa Rica, Guatemala, Nicaragua, El Salvador, Honduras, Panama and Belize. The meeting to sign the document, held just before Berlin, was attended by big business from the United States. The Declaration rehearsed the points in JI's favour. It spoke of JI being a new and additional source of funds for long-term sustainable development projects, promoting technology transfer, of contributing to cost-effective climate change policies to ensure local and global benefits. It said pilot projects in Latin America had been positive from the economic, environmental and social points of view.

Costa Rica and the United States had already signed a joint implementation agreement in June 1994 and a number of projects were underway, including a rural electrification scheme using renewable energy and reforestation. But these were not what big business had in mind. A better example was the Decin Project in the Czech Republic. Three US utilities had combined to switch the Bynov district heating plant in the city of Decin in northern Bohemia from brown coal to natural gas. The three utilities bought a 4 per cent stake in the project and in doing so provided enough capital for the fuel-switching to go ahead. Greenhouse gas emissions were cut by 31 per cent. Sulphur dioxide, which had caused severe local pollution, was virtually eliminated, along with coal ash. For JI supporters the Decin Project was a textbook case of how a small investment could produce big greenhouse savings as well as other environment benefits. It was a simple point: capital spent on JI could make a far greater contribution to an improved environment than the same money being expended at home.

In a briefing for journalists, the industry lobby said that there were many myths about JI. One of them was that JI enabled industrial countries to claim maximum credits for sorting out the easiest options for reducing greenhouse gases in the developing world. Once they had done this they would leave the developing country to sort out the more difficult problems on their own. The answer to this 'myth' was that the recipient country chose the projects it wanted help with and any credits to be accrued to the donating country were a matter of negotiation between them.

Among the list of advantages for developing countries cited were: the attraction of additional foreign capital otherwise unavailable; the transfer of modern clean technologies at a faster pace than otherwise available; training of the local population in new skills; additional employment; and sources of income for more efficient and cost-effective production.

The green lobby, however, and the majority of the developing world did not buy it. Although the attraction of new capital and efficient power plants tempted some developing countries, there were grave suspicions about motives. The developing countries believed that these schemes would give the industrialized world some leverage over them to limit their own emissions, something they wanted to avoid as long as possible. Second, the Climate Change Convention provided for technology transfer without strings. Perhaps the most powerful reason was the suspicion that JI was just a way of big countries dodging round their real responsibilities. Remember at Berlin, the United States was pushing JI but making no moves at all to force carbon dioxide cuts on its domestic market. What was needed, according to the green lobby, was not JI, but steps towards a fossil-fuel-free future at home, including investments in clean technologies in the home market and carbon taxes. They suggested that no industrialized country should be allowed to claim credits for JI unless they had implemented a 20 per cent reduction in their own carbon dioxide levels by the year 2005. Another substantial point sprung from the Czech example. Modifying existing power plants to cut emissions was a clear benefit, building new power plants, which otherwise might not be constructed at all, was surely not helpful. JI would be far better used to pay for energy-efficiency measures in people's homes in order to avoid the need to build new power plants.

Many people believe that the agreement reached in Berlin effectively killed joint implementation as a major force in climate change. It encouraged pilot projects to test the idea but ruled out any credits for industrialized countries during this phase. It promised an annual review of projects but would not consider changing the rules on JI until the year 2000. The ball had been firmly lobbed back to the industrial world. It had to prove that it was serious about reducing its own emissions if JI was to be a starter.

But as stated at the beginning of this chapter, the issue of JI is one of a number of red herrings that were allowed to muddy the waters at Berlin. In the last chapter there were figures from *Harper's* magazine on how much the fossil fuel industry in America had spent on disinformation and public relations to do with climate change. The article's author, Ross Gelbspan, also named a small band of scientific sceptics – Dr Richard S. Lindzen, Dr Pat Michaels, Dr Robert Balling, Dr Sherwood Idso and Dr S. Fred Singer, who he says have proved 'adept at draining the issue of all sense of crisis'. According to Mr Gelbspan, 'Through their frequent pronouncements in the press and on the radio and television they have helped create the illusion that the question is hopelessly mired in unknowns.

Most damaging has been their influence on decision makers.' Their contrarian view has allowed conservative Republicans to dismiss legitimate research as 'liberal claptrap' and provided the basis for the round of 1995 budget cuts to government science programmes monitoring the health of the planet.

Mr Gelbspan names Western Fuels Association, a consortium of coal suppliers and power station owners, as one of the most aggressive in rubbishing the global warming threat. They produced a video in 1991, apparently favourite viewing in the Bush White House and in OPEC country capitals, showing how increasing concentrations of carbon dioxide in the atmosphere would result in a new age of agricultural abundance. This included vast areas of the world where new grassland would replace deserts, and new forests would grow, thriving in the extra carbon dioxide. As Mr Gelbspan succinctly comments: 'Unfortunately, it overlooks the bugs.' Even a minor elevation in the earth's temperature would trigger an explosion in the earth's insect population.

But back to 1995 and the Western Fuels Association appearance at the Minnesota hearings to determine the environmental cost of coal-burning in state power plants. They hired three of the scientific sceptics – Drs Lindzen, Michaels and Balling – to give evidence as expert witnesses. At the hearings the three scientists argued between them that maximum probable warming over the next century because of a doubling in carbon dioxide emissions would amount to between 0.3°C and 1°C (32.5–34°F) and there would be no discernible sea level rise. In their annual reports the Western Fuels Association are open about their policy. They say that some of their industry members are prepared to concede the scientific premise of global warming, while arguing over policy. 'We have disagreed and do disagree with this strategy. When the controversy first erupted scientists were found who were sceptical about much of what seemed generally accepted about the potential for climate change.'

Mr Gelbspan reports: 'While the sceptics portray themselves as besieged truth-seekers fending off irresponsible environmental doomsayers, their testimony in St Paul, Minnesota revealed the source and scope of their funding for the first time.' Dr Pat Michaels, who teaches climatology at the University of Virginia, had received $115,000 over the past four years from coal and energy interests. *World Climate Review*, a quarterly he founded that routinely debunks climate concerns, was funded by Western Fuels. Over the past six years Dr Robert Balling, from Arizona State University, had received $200,000 from coal and oil interests in the United Kingdom, Germany and elsewhere. Richard Lindzen, a distinguished professor of meteorology, charged oil and coal interests $2,500 a day for his consulting services and a 1991 trip he made to testify before the Senate was paid for by Western Fuels.

Mr Gelbspan's article continues:

> The sceptics assert flatly that their science is untainted by funding. Nevertheless, in this persistent and well-funded campaign of denial they have become interchangeable ornaments on the hood of a high-powered engine of disinformation. Their dissenting opinions are amplified beyond all proportion through the media while the concerns of the dominant majority of the world's scientific establishment are marginalized.

These scientists and the political thinkers who use their material to rubbish global warming are called contrarians by journalists. They are the sort of people who in another century would have hung on to the theory that the earth was flat, long after most scientists had proved it was round. Both then and now the flat-earthers were a powerful bunch. One of the political problems that the green movement has to face in pushing its point is that in the 1990s the power of market forces reigns supreme as a political philosophy. Interfering in the market is regarded as a political crime. The fact that the market has always been interfered with and continues to be so is ignored. For example, if people really believed in market forces not a single nuclear power station would ever have been built, at least not to produce electricity. Compared with fossil fuels it has always been at least twice as expensive. The distortion in the energy market in developing countries is even more marked. Electricity is heavily subsidized because otherwise people would simply not be able to afford to turn the lights on.

What environmentalists have been trying to argue for is that the 'true' cost of building roads and generating power should be calculated before they are built. This includes the environmental cost of destruction of the countryside and the pollution from power station chimneys. The problem is that despite the best efforts of economists it is impossible to measure such things. It is like trying to put a monetary value on a poem, a piece of music or a prayer. It

205

means different things to different people. There was, for example, a tremendous row about the cost of human life after the economics section of the IPCC was drafted. The economics team of the IPCC costed a life lost in the Third World as a result of global warming at $100,000, but costed a life lost in the industrialized world at 15 times that amount. As a result of costs based on such strange monetary values, the overall findings were that the industrialized countries would suffer twice as much damage as the rest of the world because of climate change. This was despite the fact they have only 20 per cent of the world population and less than 20 per cent of the world land area. Basically, the IPCC team valued the damage done by global warming using the willingness to pay method, which means they would ask people how much they would be prepared to pay to stop, say, an animal species becoming extinct, and then calculate the value they get as the value of the species. The economists' report accepted that many more lives would be lost as a result of global warming in poorer countries than richer ones, but since they valued human life at what people were prepared to pay to avoid the risk of dying, and the inhabitants of a poor country could only afford one fifteenth of the figure for rich countries, the value put on the damage done as a result of lives being lost was very much lower.

The draft acknowledged that 85 per cent of all the land lost due to sea level rise and 78 per cent of the extra deaths would be in developing countries, but stuck to the argument that in monetary value developed countries would lose far more. Environmental campaigners were enraged. Aubrey Meyer, for the tiny green group, the Global Commons Institute, took them on in July 1995 at a meeting for policy-makers in Geneva. He questioned the concept of life being more valuable in one country than another. Surely, if ethics rather than economics ruled, then life in the developing world would have been valued identically and the sums would have worked out considerably differently. Ethics, which have so far come a poor third to politics and economics in the climate debate, did better on this occasion. Thanks to the intervention of Mr Meyer the economics chapter was heavily questioned by Cuba, Brazil, India, China and Peru. Although the calculations remain, the United Nations refused to accept the conclusions in its summary report. It says simply 'the value of life has a meaning beyond monetary value'.

That sort of intervention throws out the traditional monetary calculations and accepted economic theory on which the industrial world runs. It makes politicians of the right fear the environmental movement, which it regards as a political opponent. It also fears regulation, or what it regards as an excess of government, introduced in the name of protecting the planet.

One of the organizations that takes the threat of the environmental movement seriously is the Institute of Economic Affairs, based in Lord North Street, Westminster, London, a stone's throw from Conservative Central Office. The Institute is characterized as a right-wing think tank, in other words they provide the ideas for the radical right in the Conservative Party. One of their main themes in the 1990s has been the environment, and they set up a special unit to concentrate on it in 1993. In general, they seem to be torn between trying to prove that many environmental theories are bunk and showing that problems which do exist can be cured by market forces. The Institute has published a number of studies of the environment. In one of these, produced in 1995, *The Political Economy of Land Degradation: Pressure groups, foreign aid and the myth of man-made deserts*, the author, Julian Morris, concludes that the claim that the Sahara is expanding inexorably southwards is based on evidence of dubious scientific validity and that more rigorous scientific studies show no such desert encroachment. Having gone that far to debunk the science, Mr Morris then changes course and takes much the same road as many deep green environmentalists. For example he says: 'It is untrue that peasants and nomads, acting out of stupidity and ignorance, are the primary cause of land degradation in the developing world. In fact land degradation is caused by the actions of political entrepreneurs.' These Mr Morris labels as:

> pressure groups, foreign aid officials and bureaucrats in both developed and developing countries who call for tax money to be spent on aid. The political elite in developing countries use this 'aid' for self-aggrandizement, oppression of peasants and nomads, and subsidies to the politically important.
>
> Grandiose schemes (dams, irrigation projects, mechanized agriculture, and anti-desertification projects) enacted by state officials of developing nations, in accordance with the plans of officials at the World Bank, UNEP, and others have undermined the local institutions of land ownership, the market and customary law

which enable the peasants and nomads to adapt to changing environmental conditions.

Mr Morris says the UN Convention on Desertification recommends that decisions about land use should be devolved to local communities, which he agrees with, but baulks at the call for massive centrally funded anti-desertification projects. He says the two aims are inconsistent.

Further spending on aid is likely to be unhelpful, he adds.

> Only when individuals in the developing world are permitted to own property (especially land and water), to engage in free trade, and resolve disputes through customary law, will the circle of land degradation, poverty and hunger be broken. To speed up this process of reform, no more aid should be given to the governments of developing countries.

The study of land degradation was the Institute of Economic Affairs environment unit's fifth publication. It was no surprise that the first was called *Global Warming: Apocalypse or hot air?* Most of the scientific authorities used to debunk the IPCC version of climate change are the scientists already mentioned earlier in this chapter. Indeed the Institute stocks some of the books by these scientists, including Robert C. Balling's *The Heated Debate: Greenhouse predictions versus climate reality*. The Institute's own publication throws doubt on the assertion that there has been any measurable increase in temperature in the last 100 years. It also uses the increased cloud theory to show that the climate corrects any undue heating. The book comes to the conclusion that enhanced carbon dioxide in the atmosphere is good for crop yields and will green the deserts as a result. It concludes that no one knows yet whether the enhanced greenhouse effect will happen, and even if it does, it may prove to be beneficial. Economically the book argues that there should be no action on the issue, and that too much money is spent on researching what is an uncertainty. It does, however, true to the free market philosophy, suggest the removal of subsidies from fossil fuels.

In December 1995 the Institute of Economic Affairs continued the attack. A press release advertising the winter issue of its magazine *Economic Affairs* was entitled 'Erroneous calculations form basis for UN Assessment of Global Warming trend. Met Office withholding critical information.' Professor Patrick Michaels and Dr Paul Knappenberger set about claiming that the scientific basis for the Framework Convention on Climate Change was faulty. Clearly, however large the consensus of scientists at the IPCC, there are always going to be enough professors left over to provide ammunition for the sceptics. Perhaps their science could be described as a smokescreen for both the activities of the fossil fuel lobby and right-wing think tanks.

It is not that these groups are out on their own; the contrarians pop up all over the place. One of the best known in Britain is Richard D. North, a former environment correspondent both of the *Independent* and *The Sunday Times*. He understands the issues as well as, and probably better than, most. As one of the first environment correspondents, he was one of the first to take seriously the threats facing the planet. But then just as the rest of the world began to wake up to the problems, he started to become a sceptic. During the 1990s he has spent much time on the radio, writing columns and appearing on platforms debunking newly accepted environment theory. He expounds his own theories in his book *Life on a Modern Planet,* and is now firmly of the school that humankind's own ingenuity will enable us to feed the massively increased population, and that development is the key to defeating the environment problems the world faces.

Interestingly, he is not arguing against the precautionary approach on global warming, although he wants to slow down the mitigation action. He emphasizes the uncertainty of the science but says that the industrialized nations have a clear duty to reduce emissions. Of all the environmental threats reviewed in the book, climate change appears to bother him most. He regards the climate system, with its historic temperature 'rollercoaster', as like a fairground ride. 'Our warming influence promises to accelerate the ride, and thus be dangerous, whatever its detailed outcomes.'

Mr North also admits that it is clear that we are adding to the warming of the earth with effects we cannot be certain about. 'It is unlikely or even inconceivable that it could be a sound principle to add to a risky system a further element of riskiness, and to do so in the hope that somehow the system will accommodate the new factors and ameliorate them.'

Despite his caution about the future, at least as far as meddling with the climate is concerned, the newly sceptical Mr North suggests a wait-and-see policy on activities specifically aimed at reducing emissions. He is keen on the 'no regrets' policies of energy efficiency and other small taxes to encourage such things as public transport, which are for a general social good. But he is against the general ditching of fossil fuel power stations and other technologies until they are worn out and other as yet unproved renewable or more efficient systems are developed in their place.

Although many people dismiss think tanks and contrarians as outside the main stream it would be unwise to do so. Their political influence is very strong. Some of their theories and propaganda are very close not only to views of leading politicians of the right, but also to the opinions of writers on such papers as the London *Financial Times*. One of the writers who most influenced the White House in the late 1980s and early 1990s is Frances Cairncross of *The Economist*. When she attended climate change talks she was stunned to find that White House officials regarded her work as definitive and carried it about in their briefcases. Subsequently she has written books on the subject. The latest, *Green Inc.: A guide to business and the environment*, has been hailed by industry and politicians as a mass of common sense combined with a formidable grasp of economics. Her views on climate change can be regarded as mainstream if rather depressing. She says it will be about two decades before we can be sure how great are the threats from global warming. She wonders whether we can wait that long before taking action and concludes, like Richard North, that the 'no regrets' policies could make a significant dent in the problem while we wait and consider alternatives.

But being practical, she points to the enormous difficulties of preventing a massive rise in the carbon dioxide emissions from the developing world, particularly since that is where most of the extra two billion people will be born in the next 20 years. As an illustration of the problem she says that the average person in a developing country uses the equivalent of between one and two barrels of oil a year for fuel, the average European or Japanese, the equivalent of 10 and 30 barrels and the average American 40.

Coal, the biggest source of carbon dioxide, accounts for 70 per cent of the heat content of the world's fossil fuel reserves. She says it must remain earth's main source of fuel for electricity generation. She points out that coal will be particularly important in the world's two most populous nations: India and China. The Chinese have one third of the world's known reserves of coal. She says that in order to have any hope of persuading the Chinese to leave these stocks in the ground, any alternative source of energy would have to be cheaper than coal. At present a large conventional coal plant with no pollution controls is cheaper to build than any rival technology. Even though it might be possible to find an alternative to coal, and its rivals oil and gas, the problem is that all three are currently cheap and plentiful.

In all, she comes to the gloomy conclusion that the Climate Change Convention has little chance of making a significant difference in the world's output of fossil fuels over the next century. But she also argues that in economic terms this does not matter, at least for the next century and a half. She says it would be wise for all countries to introduce emission-reducing measures that would pay off in any case, such as raising energy prices to economically efficient levels and reducing other kinds of pollution.

Instead of adopting other targets and timetables she argues that the rational course is for us to adapt to climate change, as and when it happens.

> Most countries will be richer then, and so better able to build sea walls or develop drought-resistant plants. Money that might now be spent on curbing carbon dioxide output can be invested instead, either in preventing more damaging environmental change or in creating productive assets that will generate future income to pay for adaptation. Once climate change occurs, it will be clearer – as it now is not – what needs to be done and where.

She argues that adaptation is particularly appropriate for poor countries which should concentrate their efforts in the meantime on getting richer so that they can afford to pay for it when the time comes. She comments on her own analysis:

> Many people find such arguments unpalatable. To many environmentalists, climate change seems to be nature's revenge on humanity for economic growth. The idea of adapting to it, rather than struggling to minimize, will sound wilfully irresponsible. Yet the harsh reality is that plenty of other kinds of environmental damage deserve greater priority. Water pollution kills more

people than global warming is likely to do; soil erosion leaves more people hungry; the loss of species just as irreversible . . . The world has only so much wealth to devote to solving environmental problems. Many of these deserve greater priority than global warming.

In this chapter then we have run through the strands of thought of those to whom global warming is not an urgent problem. It is not urgent for a variety of reasons and motives. There are those who fear for their own businesses and futures if global warming is tackled seriously, others who give greater priority to their belief in the power of free market economics, right through to Frances Cairncross who believes that the climate treaty will not work, and even if it did, there are other more urgent problems to deal with.

In the next and final chapter we run through some of the political events since Berlin, the prospects for the 1997 crunch meeting in Japan, COP3, when the period beyond 2000 is addressed. The latest arguments among the politicians and gurus are reviewed. The most important questions are tackled: can there be a consensus about what measures we need to undertake? And as we venture into the twenty-first century, what nasty surprises does climate change have in store for us?

# 19

---

# The big questions
# still wait

LEST WE FORGET, THE OBJECT of the Climate Change Convention is to stabilize greenhouse gas concentrations in the atmosphere at a level that would prevent 'dangerous' anthropogenic interference with the climate system. So the big question, which politicians who signed up to the convention have failed to answer, is what constitutes 'dangerous'. They asked the scientists to answer it and they have. The IPCC, and let us remember these are the cream of the world's climate scientists, unanimously replied that the danger point has already been reached. To pull us back from the brink they ask for 60 per cent reductions on today's emissions. The politicians noted it, realized that it was beyond them, and looked around to see what they could do as a first step. The best they could manage at the time was a stabilization on 1990 levels by the year 2000. That kind of reduction is what scientists call grossly inadequate and then they go further and suggest that such small measures will make no real difference, particularly when the vast extra emissions from the developing world are taken into account. So the next big question follows from the last. If existing commitments are not adequate, and the 60 per cent reduction is not politically possible – what is the world aiming at? What level of enhanced greenhouse gases can be allowed in the atmosphere which the politicians think the world can live with, and it is possible to deliver?

This is perhaps the most important current question to which there is no answer, partly because we do not yet know at what point the climate will stabilize, and at what level the sea will stop rising. But if we make a best guess, and that is all we can hope for, then tricky calculations can be made. Let us remind ourselves again that most of the scientists' calculations of sea level rise and temperature changes are based on a doubling of carbon dioxide, that is from the pre-industrial level of 280 parts per million to 560 parts per million. In early 1996 the concentration had already risen to 360 ppm and is still rising fast. At the March 1996 meeting of climate change negotiators in Geneva there was some effort to name a maximum allowable increase for the world at 550 ppm.

So to put it in those terms, if politicians settle for that – a doubling of carbon dioxide – it is goodbye to most of the 36 AOSIS nations by the end of the twenty-first century. The populations of those nations and more than 100 million other people living on vulnerable low-lying land will have to be moved to higher ground and be found space to live, work and farm. In Geneva a cutting from a newspaper in the far-off Solomon Islands, north-east of Australia, was circulated to delegates. It reported the fact that the Carteret Islands, where 1700 people live, had just been washed over by a tidal wave. The islands' entire vegetable crop had been lost because the gardens and the soils had been washed away. Emergency rice supplies had been sent. The report said the sea level had risen 30 cm (12 in) a year since 1991 and during high tides three times a year the islands were in danger of being submerged. It predicted that the islands would be uninhabitable within five years and that the government were looking at higher islands to which the Carteret islanders could be evacuated and on which they could be resettled.

The issue of resettlement of millions people soon to face the same crisis as those in the Carterets is such a difficult political problem in a crowded world it may seem insurmountable by itself. But in Geneva and elsewhere the fossil fuel lobby and many nation states argue that targets as difficult to attain as 550 parts per million should not be attempted. As a result many believe that even stabilization at double pre-industrial carbon dioxide levels is beyond the powers of our political system. This is because of an in-built weakness of democracies that cannot plan beyond winning the next election. In other words, politicians are programmed not to think of any objective which is more than a term of office away. If the pessimists about politicians turn out to be right, the less optimistic scientists think the world is in deep trouble. In addition to the land loss due to sea level rise we are faced with mass extinctions of species as vast forests and other vegetation, unable to adapt quickly enough to climate changes, wither away and die. Our inability to grow sufficient food supplies and other horrors that follow these events would, they say, mean that we will end up entering a new dark age.

The optimists, on the other hand, think that techno-fixes and the instinct for survival will pull humanity through, although so far there is no shred of evidence for this cheerful scenario, except that we have survived previous climate changes. What proportion of the species failed to cope and died out is not known.

Whichever guess is right, or even if the answer is somewhere in between, which is where my money lies, the fact remains that the key question is not answered. In order for the Convention to work, a politically acceptable definition of what constitutes 'a dangerous level' of greenhouse gases has to be fixed. This would provide a target towards which the world can aim. Otherwise how is it possible to mould policy to meet the problem?

If the issue is not tackled and a fairly tight world total emission level fixed, based on the science, the danger is that the politicians' own targets will always be too weak. They will be influenced by their own local expediency, based on their own electoral survival, rather than on what is needed for the future. Politicians, who do not generally like finding answers to big questions, will no doubt shy away from making such a decision for as long as possible. They know that ultimately if the Convention is to work, i.e. to save the world from the worst of the enhanced greenhouse effect, a world target has to be fixed and then painful measures adopted to make it stick.

But even if politically the world is not ready to face that big question, it is trying to answer another one. Having accepted at Berlin that existing commitments are not good enough, another tricky question is automatically posed. If our existing efforts are inadequate, how are we going to reverse the trend of our currently ever-increasing greenhouse gas emissions? Answering this question before the other one has some logic, because whatever the world target is that is eventually agreed, on current trends we are never going to hit it.

Whatever we decide to do in the future we have to reverse the trend of worldwide emissions going up at an ever-accelerating rate. So as a pragmatic and practical politician might say, let us see what we can do to change the trend before we set ourselves the ultimate

## Energy use

Global energy use, in exa ($10^{18}$) joules per year, from the last century

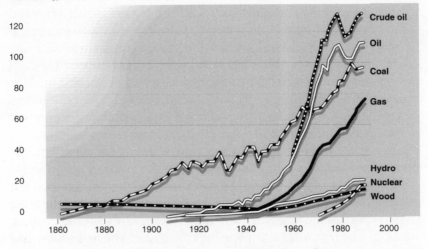

in targets and timetables. Once we can do that, then we can move on to global totals for maximum greenhouse gas concentrations and a timetable for getting them down.

By the middle of 1996 it was clear that there was still a long way to go even before countries could answer how they were going to reverse present trends. With a few notable exceptions, and the United Kingdom is one of them for reasons we shall examine shortly, the industrialized world is struggling even to reach stabilization targets. Compared with controlling CFCs, greenhouse emissions are a nightmare. Economic activity, never easy to predict, has an important bearing on emissions and corrective measures, like spending on energy efficiency, take years to have a significant impact on annual country totals.

So the relief and rather muted rejoicing at the end of the Berlin conference was principally because the Climate Convention had survived intact – and because a door had been opened for progress to be made in the future. It did not in itself make the slightest dent in the rising greenhouse gas emissions. What it did do, however, was to confirm that the ratchet effect built so carefully into the treaty before Rio was actually working. Berlin had set a timetable for new agreements to be made and sent everyone off promising to do their best to implement existing agreements to mitigate greenhouse emissions in the meantime.

To show that there was the political will for the process to con-

216

tinue, a secretariat had been created, and two committees required to keep the verification and scientific advice on track had been organized. The principal difficulties which had so nearly wrecked Berlin remained, however. In order for action against climate change to work the entire world has to be involved. But this seems a tall order with nearly 150 countries to be consulted, with widely different aspirations and at different stages of development. The power blocs, how they divide up, and some of the motives of the players have already been discussed. But even putting those on one side for a moment there are still formidable barriers to progress. So far in this book there have been limited mentions of ethics, but plenty of mentions of inequalities. As the developing world keeps reminding the industrialized world, it was not them who caused the problem; the historic releases of excess carbon dioxide are the fault of the industrialized nations. Yet as we have already discussed, it will be the developing nations which will suffer the first and worst effects of climate change. So on grounds of equity as well as ethics the industrialized world has a duty to do something about the problem.

It has become increasingly clear, however, that the industrialized nations cannot cure the problem on their own, even if they managed to create it themselves in the first place. The speed with which the Third World is developing means that emissions from the G77 countries will overtake those from the industrialized world by 2020 and will continue spiralling upwards thereafter.

It is worth re-stating again at this point that the scientific predictions we spent so much time describing in earlier chapters are based on that doubling of carbon dioxide in the atmosphere we have just discussed, and that on current trends carbon dioxide will more than double before the middle of the next century. In making their calculations about climate effects the scientists have already given politicians immense credit for actions they show few signs of taking. In other words the scientists have made the enormous assumption that the world will have managed to peg back its carbon dioxide emissions so much by about 2050 that levels will not go on rising beyond that at an alarming pace. At current rates of progress that seems highly unlikely, in which case the effects of global warming might be far worse than the scientists are currently predicting.

Leaving that on one side, a critical point that has to be addressed in negotiations is one of fairness. It is not just a question of who is creating the most emissions, both now and historically, it is also a question how to share the burden of reducing them. The argument of the Third World that it must be allowed to develop unhindered is well rehearsed, but between these nations there is already an enormous disparity in both wealth and lifestyle. Some so-called developing nations are richer and more hi-tech than some southern European countries. Similarly, some extremely sophisticated European countries and Japan have already gone a long way down the road in terms of carbon dioxide reductions. They claim that to ask them to go further would involve them in costs which would unfairly damage their competitiveness.

There are difficult issues too about how you judge how well a country is doing. How do you measure emissions? Many believe measuring carbon dioxide produced by a country per head of population is the fairest measurement of each nation's responsibility for climate change. This could be modified to include all greenhouse gases, and it could be modified again to allow credits for what are called sinks, for example vast tree-planting schemes. For many, the issue of joint implementation, claiming credits for work done in a partner country, could also play a part. There is also the European model where larger groups of countries can come to an overall target with some making larger reductions than others. Many countries are also already claiming special circumstances. This can mean anything from Australia's claim that it has certain industries that use more electricity to Norway and Sweden's belief that they have already done more than their fair share to limit emissions.

The crunch at the end of all these points is that tackling climate change remains extremely difficult. Every country appears to be able to think of any number of reasons why it has special circumstances for not taking actions its neighbours should. As we said in the Introduction, the climate change process involves tackling a traditional way of thinking. The world is currently hooked on burning cheap fossil fuels to allow continuous growth without a thought for the consequences. Despite the doubters, the disinformation and the contrarians, it is already clearer than it was in Berlin what the consequences of that tradition will be.

It is worth noting that even in the months since the third IPCC report was completed in December 1995 with its far more frightening science, even more evidence has emerged which supports the global warming theory and its potential for disastrous effects.

Before we look to the future we should review what has happened since Berlin. Perhaps the single most significant event has been the publication of the IPCC reports themselves. The scientific consensus is now stronger, but most important the scientists concluded that global warming due to human activity has already begun. That makes the whole issue more politically pressing, and makes life harder for the carbon club and the contrarians.

Another straw in the wind was the largely unreported 1995 Tokyo meeting of the World Energy Council, a 100-country body that studies future global energy strategy. Once the most fossil-fuel-orientated bunch in the world, they are now forecasting a big boost for renewables. The Council saw two fundamental challenges in its conclusions. The first was to respond to the plight of over two billion people, mainly in lower-income developing countries, who have neither electricity nor adequate access to other commercial energy.

> In consequence, they have no realistic prospects of breaking out of the vicious circle of poverty and taking the first steps towards development, a higher living standard, and the reversal of serious local environmental degradation. The second challenge was that of a path to sustainable development in the longer term. This path recognized first the inevitability of world population growth to about 8 billion by 2020 and probably 10 billion by 2050 and the imperative of economic development to provide all people with an acceptable way of life; and second, the need to come to terms honestly and fearlessly with the impacts of economic development on our environment, ranging from the local and regional to global issues such as enhanced global warming, and threats to bio-diversity. These impacts are expected to grow particularly rapidly in the south.

The World Energy Council then called on all governments to face up immediately to the actions that are required to meet these two challenges:

> Action postponed will be opportunity lost, guaranteeing that when action can no longer be avoided the ensuing costs will be higher; dislocations more severe; and the effects much less predictable, than if the appropriate actions are taken today.

Just pinch yourself at this point with a reminder that this is not an environment group talking but the world representatives of the fossil fuel lobby. The same document demands that investment decisions should be taken now to boost renewables. Depending on the decisions taken in the next 20 years renewables could be a mere 16 per cent of primary energy production by 2050, or as much as 39 per cent. The Council is advocating the removal as quickly as possible of the subsidies for fossil fuels and the addition of carbon taxes to reflect their true environmental cost. The Council also calls for 'minimum regret' action. This it describes as greater energy efficiency, greater conservation and recycling wherever appropriate; cleaner fossil fuel conversion and use, a bringing forward more quickly of economic non-fossil energy supplies; addressing the huge challenges of road transport; and the expansion of carbon sinks as well as the reduction of greenhouse gas emissions.

But you might ask, what of the continuing climate change talks themselves? What have the sherpas been up to since Berlin? At first, the news was not so good, the sherpas appeared to have fallen back into the mire which characterized the pre-COP1 meetings. As happened after Rio, the head of steam had disappeared and the sherpas had resumed their defensive positions in the slit trenches, lobbing the occasional position paper over the top at the opposition. It would be depressing to go into too much detail, but it would also be too flip to suggest there has been no progress. Only by exhaustive, and certainly exhausting, discussions about some of the issues outlined above can civil servants come up with solutions to seemingly intractable problems. The issue of fairness in adopting targets and timetables between countries has to be cracked if COP3 in 1997 is to be a success. Berlin was, after all, talking about the time frames of 2005, 2010 and 2020. The 2020 date was insisted on by the United States for the simple reason that by then the developing world will be responsible for more than half the world's emissions. In order to make its home lobby contemplate reducing American emissions the developing world will have to be seen to be playing its part, even if it is not doing so yet.

On the other hand there can be no doubt that unless the industrialized world meets its existing targets, and some far more exacting ones after the turn of the century, the developing world cannot be expected to take action. The industrial world still has to establish its credibility if it is to be able to ask seriously for the developing nations to take any action on targets, however remote.

The first meeting of the sherpas was held in July 1995 and a great deal of time was taken up with discussing possible policies and measures to cut greenhouse gas emissions. To give some idea of how long this could take, 700 possible ways of doing so have so far been identified in national plans. The second meeting in October was more productive, and just as important was better attended; a good number of developing countries could simply not afford to attend the first gathering.

The March 1996 meeting got down to more substantial business and the IPCC report showing that global warming was already underway was presented formally. There was resistance to this from the fossil fuel lobby which produced its own group of scientists and a report called *The Global Warming Debate*, in which it sought to highlight the uncertainties. However, the political meeting that followed was more positive than anything since the last days of Berlin. Targets and timetables were openly discussed by the United States for the first time. Several countries wanted 10 per cent reductions in carbon dioxide emissions by 2005 and 15–20 per cent by 2010. There were just as many disagreements, but the direct negotiations about real objectives had really begun again.

In July, at the start of the second conference of the parties (COP2) in Geneva, there was gloom about the prospects for progress. The fossil fuel lobby was there in force, with the figure of Don Pearlman constantly working away in the lobbies. He had gained the ear of the Russian delegation and warned them that falling oil prices would be bad for their already disastrous economy.

The European Union had again taken the most forward-looking position. Having reviewed the science, the EU Council of Ministers tried to set a limit on the maximum level of global warming that the climate could tolerate. They came up with a 2°C (36°F) maximum temperature increase and 550 parts per million (ppm) of carbon dioxide in the atmosphere. The figure is double the pre-industrial level of 280 ppm, and well in excess of the current 320 ppm. On present trends the world will reach 550 ppm by 2060 if only carbon dioxide is taken into account. If methane and other greenhouse gases are added into the calculation the carbon dioxide equivalent will have doubled by 2030.

Professor Bert Bolin, chairman of the IPCC, who was in Geneva, made it clear he thought that the doubling of carbon dioxide and the resultant 2°C (36°F) rise in temperature was the absolute maximum

221

extra stress that the climate could take before problems became 'critical'. In other words, he was endorsing the EU proposal. What he nor anyone else dared to do at that moment was to define this temperature rise as 'dangerous'.

This is the crunch word which appears in the climate treaty itself. All those that sign up to the treaty are supposed to be pledged to preventing the human race allowing emissions which cause 'dangerous interference' with the climate. This EU statement in Geneva, and Professor Bolin's endorsement of it, is the closest the politicians have come to defining 'dangerous'. The importance of this definition is that from it maximum emissions can be calculated.

According to the scientists, keeping the rise in the global temperature down to 2°C (36°F) means carbon dioxide emissions for the entire world must not rise above 1990 levels. In other words, if the developing world was to continue to industrialize and therefore increase its emissions then the developed world would have to start making deep cuts as soon as possible, at least from 2000. All this was too difficult for the politicians to swallow in Geneva, but provided a baseline for setting targets for 1997's COP3 in Japan. It would make for interesting negotiating sessions in the coming months.

But while these markers were important, the three days of political negotiations in Geneva proved to be dramatic because of a complete change in tactics by the United States. As has been noted, the Americans had already begun to shift their ground by being prepared to talk about targets and timetables. Timothy Wirth, undersecretary of state for Global Affairs, who had such a hard time in Berlin, made a dramatic intervention on the last Wednesday of the conference. First of all he repudiated the fossil fuel lobby by making it absolutely clear that the work of the IPCC had the 100 per cent backing of the US government.

In a single political swipe he attacked the 'naysayers and special interests bent on belittling, attacking and obfuscating climate change science'. Months of work by Don Pearlman and his allies in the Global Climate Coalition, in undermining the science by planting articles in such papers as the *New York Times* and the *Wall Street Journal*, was swept away. Second, and just as important, was Mr Wirth's more unexpected endorsement of the idea of legally binding targets and timetables for reducing greenhouse gas emissions into the atmosphere. The big surprise was the phrase 'legally binding'; critical, because at present the existing targets of getting emissions

222

down to 1990 levels by 2000 are binding in honour only. The new proposal means that countries which fail to meet their targets can be taken to the International Court of Justice or more potentially damaging, face trade sanctions from the rest of the world.

This announcement sent shock waves through the conference and was rapidly followed by a powerful speech from John Gummer for the UK. The conference was taking place the Palais des Nations, the huge building set up to house the League of Nations in the 1930s. Mr Gummer urged delegates not to let the world down as the League had done before the Second World War when they stood by and allowed powerful interest groups to destroy the peace. He said climate change was potentially as dangerous to the human race as the Second World War itself.

He said everyone in the giant conference hall would feel the effects of climate change: 'No one in this hall is so old that they will not feel the effects of climate change unless they get run over by a Geneva tram before the end of this conference.'

Mr Gummer, who got the loudest cheer of any of the 50 politicians giving speeches at the conference, directly criticized the Australians. Having once been among the greenest of nations, Australia had totally reversed its position and was what Mr Gummer described as a 'back marker'. He accused their government of 'putting exports of coal to Japan as more important than the future of the next generation of Australians'.

The US and EU combined to give the conference the political boost it needed. An all-night session followed in which a ministerial declaration was hammered out. It noted the vital EU figures on climate limits, without actually adopting them, but accepted the American idea of legally binding targets. No targets were set, however. That was left to Japan in 1997.

Despite the fact that the declaration was merely a politican statement, some countries could not accept it. Australia became a pariah of the industrialized world by entering a reservation. Fourteen oil countries led by Saudi Arabia, and including Russia, also objected. It was the first time there had not been a consensus at climate talks but 150 countries, including the most powerful economies, had signed up. The ratchet effect was working again.

This is a remarkable turnround in seven years. Let me illustrate with a couple of quotes from the period. Margaret Thatcher in 1989 on Britain's stabilization target then set at 2005: 'This is a very demanding target requiring significant adjustments to the economy. It will mean more efficient power stations, cars using less fuel, better insulated homes and better management of energy in general.' David Trippier, then minister of environment, in July 1990: 'It will cost governments a fortune and the electorate pain and anguish. They will have to suffer in order to save the planet. Britain faces a bumpy and unpredictable ride.'

Both sets of remarks have some truth in them, but the problems were nothing to do with global warming. Figures released in December 1995 showed that the UK would undershoot its 2000 target by between 4 and 8 per cent, without making any further effort. This had nothing to do with Mr Gummer or anyone else caring about climate change, it was a spin-off from other government policies. In fact nearly all the measures the government had said it would take in the early 1990s to reduce emissions had either not got off the ground at all or had fallen well below target. What had made all the difference was a massive reduction in the amount of electricity being produced by coal. It was falling from a 68 per cent market share in 1990 to a projected 25 per cent in the year 2000. The nuclear industry, plus imports from France, had increased its share of the market from 23 per cent to 36 per cent, but the most important factor was the switch of some 32 per cent of production capacity from coal to gas. This is significant because gas produces about 50 per cent more electricity for each tonne of carbon dioxide released into the atmosphere.

Mr Gummer, of course, claims that this dash for gas, as it is called, is part of the government's climate change strategy, but it has a lot more to do with the privatization of the electricity industry and the closure of the coal mines for political reasons.

Just an additional point here: the attitude of the UK and other governments may alter with the political climate. It is hard to tell what effect a change of government may have because the Labour Party in the UK has not been in power while the environment has been such a strong issue. However, the Labour leader, Tony Blair, has personally made commitments on the environment, the most important on this subject being a carbon dioxide reduction of 20 per cent by 2010. Coupled with the phasing out of nuclear power, this equates to a much tougher target than the Conservatives were prepared to offer. This Labour Party commitment might be important at the 1997 Conference of the Parties.

Even more depends on the fate of President Clinton and the swings in the political climate in the United States. A crucial issue is whether the Democrats feel strong enough to carry public opinion with them and take on the Republicans and fossil fuel lobby on the issue.

(One point of optimism here, outside the political questions, is that carbon dioxide emissions go up and down for reasons totally unrelated to the efforts of politicians trying to make them change. As we have just discussed, despite all the predictions to the contrary, Britain's carbon dioxide emissions are going down, without the government really trying. In fact because of the former predominance of heavy industry, emissions are currently below those of the 1960s and 1970s. The same is true of a number of former Eastern bloc countries. In fact, in 1993 for the whole of the industrialized world emissions were only 6 per cent higher than 20 years ago. This is still far too high but it does not necessarily mean that emissions need go up for ever.)

But to get back to the UK's rather hypocritical claims. Figures released in December 1995 from the Building Research Establishment showed that greater energy efficiency could cut an estimated 11.5–17 million tonnes a year from the total carbon emitted from energy use in homes. Twenty efficiency measures were evaluated, including insulation, double glazing, draught proofing, the use of condensing boilers, low-energy lights, and switching to efficient electrical appliances and gas cookers. The point about all these calculations was that they were all 'no regret' measures, in other words the capital cost of introducing them would be paid back in a relatively short time.

However, to get a national plan working would require political leadership, and in Britain, where existing targets are already being met, there was no wish in the Conservative government to push even no regrets measures. The Building Research Establishment figures show that in theory at any rate Britain could have reached even the Toronto Target of 20 per cent reductions by 2005 if it had tried to do so by introducing an energy efficiency programme immediately after the Earth Summit as had been promised. The Labour Party has, however, latched on to this 'win win' principle and has promised 50,000 new jobs in the UK with an energy efficiency programme towards its 2010 target – in effect Toronto postponed five years.

This does reinforce the point often made by environment groups that tough targets and timetables are essential if governments are actually going to do something other than drift.

A row which erupted just before the sherpas' meeting in March 1996 seems set to dominate much of the economic and therefore political thinking running up to the 1997 meeting. Tom Wigley, one of the best-known names in climate science, appeared in a paper published in the magazine *Nature* to back the fossil fuel lobby's campaign that action on climate change was premature. Mr Wigley, who works at the University Corporation for Atmospheric Research in Boulder, Colorado, and was formerly at the University of East Anglia, has written the paper with two other scientists, both of whom had the financial support of the United States' energy industry. The basic point they are making is that allowing existing coal, oil and gas stations to run to the end of their current economic life makes financial sense. This allows new technologies to be fully developed so that when they are replaced the maximum savings in emissions are made. In other words, large emissions now and much smaller ones later, rather than installing intermediate technology which would run for longer periods. The argument goes that at the end of the twenty-first century the aggregate of total emissions would be the same by both methods.

This argument has appalled many scientists and the environment lobby. One of its greatest weaknesses is that it makes the assumption that carbon dioxide can be allowed to rise quickly to 550 parts per million in the atmosphere. Mr Wigley is presupposing that doubling of carbon dioxide or more is all right and that it can be cut back later, displaying a faith in the ability of politicians to control events which so far seems unjustified. In another sense he is also anticipating that we can manage the sea level rise and other problems that this level of emissions will cause. Experience may well prove his optimism misplaced on both counts, by which time it will be too late.

In his defence, Mr Wigley has since gone on record to say that he was not advocating no action, merely trying to add economic and political reality in tackling the issue. He wants the politicians to fix an upper limit to carbon dioxide concentrations as soon as possible because he accepts the issue has to be addressed.

Michael Grubb, head of the environmental programme at the Royal Institute of International Affairs in London, who reviewed Mr Wigley's paper, entered the controversy. He said that publication went ahead despite some of his reservations. He described some of the economics in the paper as 'meaningless'. For example, electricity generation stock is being replaced all the time and so each new plant should be less carbon intensive than the last, adopting the best technology at each new opportunity to do so. He quotes the World Energy Council's point that 'action postponed will be opportunity lost, etc.', already discussed earlier in this chapter. Mr Grubb also makes the point that unless there is pressure for change now, the new technologies necessary to make climate mitigation work will not be given priority and developed at the speed needed.

Clearly this row will run and run. Having lost the argument over the science, although they are still not admitting it, the fossil fuel lobby is now falling back on the economics. If they can find scientists and economists who can argue that doing nothing for now is a good thing then they will find eager listeners among some politicians.

So what are the realistic prospects for tackling the enhanced greenhouse effect? The scientists remain solid that we need at least 60 per cent reductions on current emissions to cure the problem we have already created, and that the problem will rapidly get worse if we continue with business as usual. The politicians who are supposed to deal with the crisis are so far not doing very well. In the industrialized world governments are struggling to hold emissions down to current levels and everyone agrees that the developing world has other priorities and will not have to make cuts, at least not yet.

But the process to bring these two positions closer together is on track. The ratchet effect of the Climate Change Convention is still working, albeit slowly. An advantage is that the science is getting ever more certain. With each extreme weather event the evidence that we are approaching a crisis grows all the time. Perhaps we can hope that enough of these events happen in the run-up to the 1997 talks to sharpen the public's imagination and the politicians' resolve. One of the key factors remains the attitude of the United States. As stated in reporting the Berlin meeting, the US delegation came good in the last hours and saved the conference from disaster with a deal. In the final IPCC negotiations in late 1995 the Americans did not back off on the science despite the intense activities of the fossil fuel lobby. The scientific case therefore remains robust, pushing the politics forward at the 1996 COP2 talks in Geneva.

The best guess then is that by 1997 politicians will be under considerable pressure to come up with a deal that satisfies public expectations, which will mean the setting of tough targets and timetables beyond 2000. Mr Gummer will almost certainly not be there since the UK general election will be before COP3 in Japan, but his 5–10 per cent cuts by 2010 are the sort of figures that will be on the table. It will be disappointing if the politicians cannot do better than that. More likely there will be a range of options up for discussion asking the world to make promises binding up to 2020. But with the ratchet effect this could rapidly improve at further COP meetings as the global crisis looms closer and the evidence mounts.

Merylyn McKenzie Hedger, from the World Wide Fund for Nature, who attends all these meetings, commented:

> There are nightmare scenarios in global warming, but nothing ever quite works out as we think it will. We make assumptions about population growth, human health issues and AIDS and then 10 years later things might look completely different. Fifty years ago almost no one had a car, in fifty years that could be true again. Society could change as much in the coming half century as it has in the last, in fact it would be surprising if it did not. At the moment the fossil fuel lobby is the most powerful in the world, but in reality they are industries on death row. In 10 years' time no one in their right mind will be investing in fossil fuel.

But there can be no doubt that the problem is urgent, and that is not currently reflected in the action being taken. As Adam Markham, the head of the WWF climate change campaign, put it succinctly: 'If we do not begin to get this right before the end of this century there will be no turning back in 30 years' time.'

# Bibliography

Balling, Robert C. *The Heated Debate: Greenhouse predictions versus climate reality*. Pacific Research Institute for Public Policy, 1992.

Bate, Roger and Morris, Julian *Global Warming: Apocalypse or hot air?* Institute of Economic Affairs, London, 1994.

Brenton, Tony *The Greening of Machiavelli, The Evolution of International Environmental Politics*. Royal Institute of International Affairs, London, 1994.

Cairncross, Frances *Green Inc.: A guide to business and the environment*. Earthscan, London, 1995.

Climate Network Europe *Joint Implementation from a European NGO Perspective*. Climate Network Europe, 1994.

Department of the Environment *The Potential Effects of Climate Change on the United Kingdom*. HMSO, London, 1991.

Grainger, Alan *The Threatening Desert: Controlling desertification*. Earthscan, London, 1990.

Greenpeace *The Climate Time Bomb*. Greenpeace International, Amsterdam, 1994.

Grubb, Michael and Anderson, Dean *The Emerging International Regime for Climate Change: Structures and options after Berlin*. Royal Institute of International Affairs, London, 1995.

Houghton, John *Global Warming: The complete briefing*. Lion Publishing, London, 1994.

Hurrell, Andrew and Kingsbury, Benedict *The International Politics of the Environment*. Clarendon Press, Oxford, 1992.

IPCC (United Nations Intergovernmental Panel on Climate Change) *Climate Change: The IPCC Scientific Assessment*. Cambridge University Press, 1990.

IPCC I *The Science of Climate Change*. Cambridge University Press, 1996.

IPCC II *Economic and Social Dimensions of Climate Change*. Cambridge University Press, 1996.

IPCC III *Impacts, Adaptations and Mitigation of Climate Change: Scientific Technical Analysis*. Cambridge University Press, 1996.

*Medicine and War*, volume 11(4), October–December 1995. Frank Cass, London.

Morris, Julian *The Political Economy of Land Degradation: Pressure groups, foreign aid and the myth of man-made deserts*. Institute of Economic Affairs, London, 1995.

North, Richard D. *Life on a Modern Planet: A manifesto for progress*. Manchester University Press, 1995.

Radford, Tim *The Crisis of Life on Earth*. Thorsons, London, 1990.

Rowlands, Ian H. *The Politics of Global Atmospheric Change*. Manchester University Press, 1995.

Scheer, Hermann *A Solar Manifesto*. James & James, London, 1994.

Stockholm Environment Institute *Confronting Climate Change: Risks, implications, responses*. Cambridge University Press, 1992.

Tooley, Michael and Jelgersma, Saskia *Impacts of Sea Level Rise on European Coastal Lowlands*. Institute of British Geographers, London, 1992.

US National Academy of Sciences *Understanding Climate Change: A program for action*. National Academy of Sciences, Washington DC, 1975.

World Commission on Environment and Development *Our Common Future* (The Brundtland Report). Oxford University Press, 1987.

World Wide Fund for Nature *Some Like It Hot*. WWF, London, 1993.

# Index

Houghton, Sir John 35, 60, 86, 94–5
hurricanes 78–9, 187
hydrochlorofluorocarbons (HCFCs) 61
hydroelectricity 139–40
hydrofluorocarbons (HFCs) 61, 62

ice, melting 75–6, 85, 87–8, 89–92
ice ages 52, 53, 71, 89
ice sheets 87, 88–92
icebergs 88, 89, 90
India 169–70, 211
industry: energy efficiency 137–8,
    141–2; water shortages 116
insect pests 128–31, 204
Institute of Economic Affairs 207–9
insurance companies 104, 182,
    185–8, 190, 191, 193–4
Intergovernmental Negotiating
    Committee (INC) 37, 147–8
Intergovernmental Panel on Climate
    Change (IPCC) 4–5, 18, 28, 34,
    36, 60, 76–7, 83, 108, 114, 115,
    131–2, 142, 177–8, 179–80,
    188, 206, 208, 213, 218–19
irrigation 116
islands 95–7, 190

Japan 29–30, 98, 105, 162, 201
joint implementation (JI), reduction
    of carbon dioxide emissions 155,
    172, 199–203, 218
Joly, Carlos 192
Jordan 110
JUSCANZ 161–2

Kelly, Bob 191
Knappenberger, Dr Paul 209
Kohl, Dr Helmut 151–6, 157, 158,
    159, 200–1

Labour Party 224, 225
lakes 73–4
landslides 122
Leggett, Dr Jeremy 182, 184, 185,
    186, 187–8, 190–1, 195, 196
leishmaniasis 130–1
Lindzen, Dr Richard 203, 204, 205
'Little Ice Age' 70–1
London 102–3, 185, 186, 190
Ludvigsen, Philip 191
Lyme disease 130

Machiavelli, Niccolo 175
Mahlman, Dr Jerry 191–2
maize 112–13

Major, John 42
malaria 129, 130
Maldives 96–7, 104
malnutrition 4, 131–2
market forces 205–6
Markham, Adam 112, 176, 196–7,
    198, 228
Mars 51–2
Merkel, Angela 151, 166, 167, 169
methane 51, 55–7, 143–4
Meyer, Aubrey 206
Miami 99
Michaels, Dr Pat 203, 204, 209
Middle East 36, 37, 106, 107,
    108, 110
migration 9, 136
minority reports 179–80
Mississippi delta 99
monsoons 75, 100
Montreal Protocol 24–7, 30, 44,
    59, 62, 171
Morris, Julian 207–8
mosquitoes 128, 129–30
mountains 120–2
mudslides 122

National Academy of Sciences 17, 34
Netherlands 98
night-time temperatures 70, 72, 75
Nile, River 99, 108–10
El Niño 77, 127–8
nitrogen 84
nitrous oxide 51, 57–8
North, Richard D. 209–10
nuclear power 139, 201, 205, 224
Nutter, Frank 189

oceans 75–7, 80–2; absorption of
    carbon dioxide 80–2; currents
    75–7, 80; sea level rises 4, 19–20,
    85–92, 94–106, 183–4, 190, 214
oil 54, 138, 139, 162, 165–6, 176,
    177–8, 195, 210
OPEC 138, 158, 162, 204
ozone 51, 59
ozone layer 6, 21, 23–7, 58–62

Pacific Ocean 77, 95
Pearlman, Donald H. 175, 176–8,
    189, 196, 221, 222
permafrost 57, 123–4
Philippines 158–9, 201
photovoltaic solar cells 141, 191
Pinatubo, Mount 67, 147, 158
plague 130